RAINFORESTS

A GUIDE TO RESEARCH AND TOURIST FACILITIES AT SELECTED TROPICAL FOREST SITES IN CENTRAL AND SOUTH AMERICA

James L. Castner

With A Foreword By Peter H. Raven

First Edition

FELINE PRESS
P.O. Box 7219
Gainesville, Florida
USA   32605

RAINFORESTS: A GUIDE TO RESEARCH AND TOURIST FACILITIES AT SELECTED TROPICAL FOREST SITES IN CENTRAL AND SOUTH AMERICA

By James L. Castner

Published By:

FELINE PRESS
P.O. Box 7219
Gainesville, FL 32605, USA

All rights reserved. No part of this book may be reproduced or transmitted in any form or by any means, electronic or mechanical, including photocopying, recording or by any information storage and retrieval system without written permission from the author, except for the inclusion of brief quotations in a review.

Copyright © 1990 by James L. Castner
  First Printing 1990
  Printed in the United States of America

ISBN 0-9625150-2-7

Library of Congress
Catalog Card Number:
89-81847

To Janice,

Whose endless patience and infinite capacity to love and find the good in people, brought happiness into my life. Thank you for everything.

ACKNOWLEDGEMENTS

Many people have contributed their kind assistance and thoughtful comments to this project and manuscript. First and foremost, I would like to thank Peter Raven for several years of continuous encouragement and support. I am indebted to him for taking time from his extremely busy schedule to write the Foreword to this book. I would also like to give special thanks to Al Gentry, whose comments allowed me to provide a much more thorough and accurate evaluation of the sites and forest types discussed. My sincere gratitude goes out to the following friends and colleagues who generously reviewed all or portions of the manuscript: Archie Carr III, David Carr, Cal Dodson, Tom Emmel, Cristian Feuillet, Howard Frank, Jeffrey Froke, Maurice LeFranc, Tom Lovejoy, Lucinda McDade, Roy McDiarmid, Geoffrey Mellor, Scott Mori, Charlie Munn III, Norman Myers, David Neill, Ted Parker III, Ghillean Prance, Ira Rubinoff, Margaret Schaeffer, Don Stone, and Stuart Strahl. Corrections to Spanish vocabulary were provided by Miriam Guizani, Charlie Munn III, Enrique Ortiz, and Helena Puche.

Many people also provided logistical assistance during my travels, both in the form of information and actual transportation; I would like to thank Alberto Fernandez Badillo, Peter Baker, Francisco Cerda, Peter Feinsinger, Marty Crump, Boris Gomez, Jean Jacques de Granville, Peter Jenson, Phil Lounibos, Julia Meredith, Scott Mori, Charlie Munn III, Giovani Onore, Doctor and Margarita Schenkel, Eric Schwartz, and Norris Williams.

The physical process of typing, correcting, revising, and editing a manuscript is a dull and dreary job. I would like to acknowledge the tremendous job done in this area by my wife, Janice Castner, who generously made time to help me in addition to her own full-time career. Her efforts in the production process were complemented by the competent work of Jill Harper and Jane Long. I would also like to thank Elizabeth Talmage whose typesetting expertise allowed the final manuscript to be translated into professional copy. Both Janice Castner and Elizabeth Talmage contributed significantly to the final design of this book. I would also like to thank Marjorie Niblack for her graphics work in creating the map illustrations.

Everyone needs a confidant, and I am no different. Many times it was only my wife's support and confidence that kept me going.

Discussions between friends and fellow biologists can also be very important. For years before I actively began to work on this project, I would deliberate at length over its various aspects with one friend who knew how much writing this book meant to me. For all the bull sessions, for all the times he listened to me go on and on, and mainly for his friendship, I would like to thank my buddy, Gary Fritz.

Finally, I thank my parents, who have enabled me to receive an excellent education and encouraged me to excel in scientific endeavors. Their support from childhood to present, along with the understanding and encouragement shown by all my other family members, have made it possible for me to work in the field of my choice and accomplish many of my goals.

WARNING - DISCLAIMER

The author/publisher has made a concerted effort to provide accurate and reliable information. However, there is no guarantee that conditions, costs, or other subjects described in this book have remained unchanged since the time when they were evaluated. Radical changes in governments, increasing population pressures, and faltering economies have all contributed to cause various degrees of instability in the countries of Latin America.

The author/publisher further makes no claims regarding the physical safety of any person traveling to locations discussed in this book. People travel at their own risk. Factors such as disease, snakebite, terrorists, and anything else capable of causing bodily harm or physical hardship should be considered beforehand and remain the responsibility of the individual.

Furthermore, the author/publisher can not be held responsible or liable for any misinformation or erroneous data or statements that may appear in this book. The strict review process and checks for accuracy have hopefully eliminated all but the most minor of these errors.

Finally, the mention of any company or organization by the author/publisher should not be considered as an endorsement, unless specifically stated as such.

TABLE OF CONTENTS

List of Photographs xiii
List of Drawings xvi
List of Maps xviii
Foreword xix
Preface xxv
Introduction xxix

Chapter 1 Countries, Sites, and Facilities

A. Peru

General Information 1
Map 2
Introduction 3
Manu Lodge 4
Explorer's Inn 10
Cuzco Amazonico 16
Tambo Lodge 21
Explorama Lodge 21
Explorama Inn 26
Explornapo Camp 31
Facilities In The Iquitos Area 37
Lima 41
Cuzco 42
Comments On Peru 44
Books 45
Maps 47
Tourist Information Sources 47
Conservation Organizations 48
Scientific Organizations/Institutions 49

B. Ecuador

General Information 51
Map 52
Introduction 53
Tinalandia 54
Rio Palenque Science Center 60

La Selva 65
Jatun Sacha Biological Station 70
Aliñahui Cabins 76
Maquipucuna Tropical Reserve 78
Rio Guajalito Reserve 79
Quito 80
Tena 81
Comments On Ecuador 82
Books 82
Maps 84
Tourist Information Sources/Travel Agencies 84
Conservation Organizations 85
Scientific Organizations/Institutions 86

C. French Guiana

General Information 89
Map 90
Introduction 91
Placer Trésor 92
Saül 95
Lassort Lodge 99
Cayenne 102
Comments On French Guiana 103
Books 103
Maps 104
Tourist Information Sources 105
Conservation Organizations 105
Scientific Organizations/Institutions 106

D. Venezuela

General Information 109
Map 110
Introduction 111
Rancho Grande 112
Canaima 118
Camp Canaima 119
Camp Ucaima 124
Camturama Amazonas Resort 125
Manaka Lodge 129
Hato Piñero 130

Caracas 130
Comments On Venezuela 132
Books 132
Maps 137
Tourist Information Sources 137
Tourist/Travel Agencies 138
Conservation Organizations 139
Scientific Organizations/Institutions 140

E. Trinidad

General Information 143
Map 144
Introduction 145
Asa Wright Nature Centre And Lodge 146
Simla Research Station 151
Victoria Regia Research Station 157
Comments On Trinidad 157
Books 157
Maps 159
Tourist Information Sources 160
Scientific Organizations/Institutions 161

F. Costa Rica

General Information 163
Map 164
Introduction 165
Monteverde Cloud Forest Reserve 166
La Selva Biological Station 171
Selva Verde Lodge 178
Rara Avis 182
Green Turtle Research Station 183
Tortuga Lodge 185
Marenco Biological Station 185
San Jose 186
Comments On Costa Rica 187
Books 187
Maps 189
Tourist Information Sources/Travel Agencies 189
Conservation Organizations 190
Scientific Organizations/Institutions 191

G. Panama

General Information	195
Map	196
Introduction	197
Barro Colorado Island (Smithsonian Tropical Research Institute)	198
Isla Maje Scientific Reserve	203
Nusagandi	206
Parque Nacional Soberania (Soberania National Park)	207
Summit Botanical Gardens	208
Panama City	208
Comments On Panama	208
Books	209
Maps	211
Tourist Information Sources/Tourist Agencies	211
Conservation Organizations	211
Scientific Organizations/Institutions	212

Chapter 2 Rainforest Information Sources

Introduction	215
Natural History And Ecological Books For Laymen	218
Adventure, Travel, And Exploration Books (1950-Present)	230
Adventure, Travel, And Exploration Books (Pre-1950)	234
Books For Biologists	242
Flora And Fauna Guides	260
Birds	260
Butterflies	263
Mammals	265
Reptiles And Amphibians	266
Floras	268
Environmental, Political, And Miscellaneous Titles	271
Travel Guides And Regional Titles	278
Statistical And Reference Books/Directories	280
Magazines For Naturalists	283
Journals For Biologists	286
Maps	290
Organizations	291
American Forestry Association	291
Basic Foundation	292
Bat Conservation International	292
Better World Society	293

Canyon Explorers Club 294
Conservation International 295
Environmental Defense Fund 295
Food and Agriculture Organization 296
Global Tomorrow Coalition 296
Greenpeace 297
International Rivers Network 297
International Society For The
     Preservation Of The Tropical Rainforest 298
Learning Alliance 299
National Audubon Society 299
Natural Resources Defense Council 300
Nature Conservancy 301
Partners of the Americas 302
Peace Corps 302
Rainforest Action Network 303
Rainforest Alliance 304
Rainforest Health Alliance 305
RARE Center For Tropical Bird Conservation 305
Sierra Club 306
Smithsonian Institution 307
South American Explorers Club 307
Summer Institute of Linguistics 309
Tropical Forests Forever 309
United Nations Information Centre 310
Wildlife Conservation International 310
World Forestry Center 311
World Resources Institute 312
World Wildlife Fund 313

Chapter 3 'Hands-On' Organizations

Introduction 317
American Ornithologists' Union 317
Caribbean Conservation Corporation 317
Earthwatch 319
Foundation For Field Research 320
International Research Expeditions 321
Monteverde Institute 321
Organization For Tropical Studies 322
School For Field Studies 324

Smithsonian Institution 325
Smithsonian Research Expeditions 326
University Research Expeditions Program 327

Chapter 4 Sources of Funding

Introduction 331
Selected List Of Organizations With Programs
   Potentially Supporting Research In The Tropics 333
     Agency For International Development 333
     American Association Of University Women
      Education Foundation 334
     Earthwatch/Center For Field Research 335
     Explorers Club 335
     Foundation For Field Research 336
     Fullbright Scholar Program 336
     Garden Club Of America 337
     Man And The Biosphere Young Scientists
      Research Grants 338
     McNamara Fellowships Program 338
     National Geographic Society 339
     National Science Foundation 339
     National Wildlife Federation 340
     (Jessie Smith) Noyes Foundation 341
     Organization For Tropical Studies 341
     Roger Tory Peterson Institute 342
     School For Field Studies 342
     Sigma Xi 343
     Smithsonian Institution 343
     The Tides Foundation 344
     University Research Expeditions Program 345
     Whitehall Foundation 346
     Wildlife Conservation International 346
     World Wildlife Fund 347

Appendix A Traveling And Travel Agencies 351
Appendix B Specialized Vocabulary (English/Spanish) .. 361
Appendix C Tropical Biologists 367
Appendix D Selected Public Zoos and Botanical Gardens . 375
Message From The Author 379
About The Author 380

List of Photographs

Page

6	The Manu Lodge located in the Manu Biosphere Reserve. (In the Department of Madre de Dios, Peru)
12	The Explorer's Inn lodge building at the Tambopata Wildlife Reserve. (Near Puerto Maldonado, Peru)
13	A guest bungalow at the Explorer's Inn. (Near Puerto Maldonado, Peru)
18	Guest cabins at the Cuzco Amazonico Lodge. (Near Puerto Maldonado, Peru)
24	An elevated guest bungalow at the Explorama Lodge. (Near Iquitos, Peru)
29	A guest cabin at the Explorama Inn. (Near Iquitos, Peru)
34	Sleeping quarters at the Explornapo Camp. (Near Iquitos, Peru)
34	The open-air lodge at the Explornapo Camp. (Near Iquitos, Peru)
56	Guest cabin at Tinalandia. (Near Santo Domingo de los Colorados, Ecuador)
57	Hotel dining room at Tinalandia. (Near Santo Domingo de los Colorados, Ecuador)
63	The biological station at the Rio Palenque Science Center. (Near Santo Domingo de los Colorados and Quevedo, Ecuador)
68	Guest cabins at La Selva. (Near Coca, Ecuador)
68	Main lodge building (R) and dining hall (L) at La Selva. (Near Coca, Ecuador)

Page

74	Sleeping quarters at the Jatun Sacha Biological Station. (Near Puerto Misahualli, Ecuador)
74	Laboratory and dining area of the Jatun Sacha Biological Station. (Near Puerto Misahualli, Ecuador)
78	Elevated guest cabin with detached groundlevel bath at the Cabañas Aliñahui. (Near Puerto Misahualli, Ecuador)
94	Hammock house and sleeping quarters at Placer Trésor. (Near Roura, French Guiana)
98	ORSTOM facilities at Saül. (French Guiana)
102	Hammock house and sleeping quarters at the Lassort Lodge. (Near Maripasoula, French Guiana)
114	Rancho Grande Biological Station in Henri Pittier National Park. (Near Maracay, Venezuela)
116	Rancho Grande Biological Station in Henri Pittier National Park. (Near Maracay, Venezuela)
121	Hillside guest cabin at Camp Canaima. (Canaima, Venezuela)
122	*Tepuis* and waterfalls visible from the dining hall of Camp Canaima. (Canaima, Venezuela)
128	Modern guest cabins at the Camturama Amazonas Resort. (Near Puerto Ayacucho, Venezuela)
148	Rear view showing the screened porch attached to the back of the estate house at the Asa Wright Nature Centre and Lodge. (Near Arima, Trinidad)
155	Main building of the Simla Research Station (formerly the William Beebe Tropical Research Station). (Near Arima, Trinidad)

Page	
168	Field station maintained by the Tropical Science Center at the Monteverde Cloud Forest Reserve. (Near Monteverde, Costa Rica)
169	The Pension Flor Mar, a small rooming house approximately 2km from the Monteverde Cloud Forest Reserve. (Monteverde, Costa Rica)
174	Dining hall and administrative offices of the La Selva Field Station operated by the Organization for Tropical Studies. (Near Puerto Viejo, Costa Rica)
174	Newly constructed modern laboratory at the OTS La Selva Field Station. (Near Puerto Viejo, Costa Rica)
175	Long-term researcher's cabin at the OTS La Selva Field Station. (Near Puerto Viejo, Costa Rica)
175	Student or field class bungalow at the OTS La Selva Field Station. (Near Puerto Viejo, Costa Rica)
181	Rear view of the Selva Verde Lodge. (Chilamate, Costa Rica)
200	Dining hall (ground floor) and living quarters (second floor) at the Smithsonian Tropical Reserach Institute on Barro Colorado Island. (Panama)
201	Laboratory building at the Smithsonian Tropical Research Institute on Barro Colorado Island. (Panama)

List of Drawings

Page	
xxviii	Traveler's tree
xxxvi	Jaguar
50	Galapagos giant tortoises
87	Tamandua
88	Anaconda
107	Pterodactyls
108	Oilbirds
141	South American poison frog
142	Hercules beetles
162	Quetzals
192	A species of toucan
193	Bird spider
194	South American iguana
213	Horseshoe bat
214	India-rubber tree and banyan tree
314	A type of fruit bat
315	Black caimans
316	A species of archerfish
328	A species of aracari
329	Racket-tailed hummingbird

Page

330	Brazilian horned frog
348	Boa constrictor
349	A species of softshell turtle
350	A type of spider monkey
360	Harlequin beetle
366	King vulture
373	Matamata
374	Emerald tree boa
377	Silky anteater
378	Lantern fly
379	Leaf insect
380	Giant anteater

List of Maps

xxxiv	Central America
xxxv	South America
2	Peru
52	Ecuador
90	French Guiana
110	Venezuela
144	Trinidad
164	Costa Rica
196	Panama

FOREWORD

The opportunity to visit and to enjoy the tropics is greater now than it ever has been in the past, thanks to the provision of reasonably comfortable facilities, conveniently located, throughout the principal tropical regions of the world. The research and tourist facilities of Central and South America are clearly described in the present volume, which affords a key to going and experiencing for yourself the exuberance of tropical life in all of its beauty and intricacy. The extensive attention given to tropical forests by the media in recent years ought to provide an impetus for learning something about them, and the opportunities for research in these forests are both extensive and of critical importance for human survival. If this book stimulates more people to visit the tropics and to conduct research there, it should be judged an enormous success.

A second, and deeply troublesome reason that the opportunity to visit and to conduct research in the tropics is timely, is that opportunities to do so will be much more limited in the future. The forests themselves are being lost, with potentially tragic consequences not only for the people who live in and around them, but for all humans throughout the globe. The atmospheric consequences of forest destruction in the intensification of the greenhouse effect have been widely discussed, but their role in generating local rainfall and in regulating the fertility of the soils and the supplies of water in impoverished nations have been a lesser source of concern, at least among the wealthy people who live in industrialized nations.

In fact, the destruction of tropical forests is proceeding so rapidly that within a very few decades, few nations will still possess more than a tiny fraction of the riches that they originally possessed. This tragic loss--an enormous blow against the chances of our attaining global stability--comes about for many reasons, but these may perhaps be summarized best by stating that the forms of government that humans have evolved over thousands of years have not yet provided adequate ways to solve the problems of the present.

The relatively gradual increase in global population, from a postulated 250 million people at the time of Christ to ten times that many in 1950, has become explosive at present. This growth, which has doubled the number of humans on earth over the past 35 years and will add another billion during the 1990s, threatens to outstrip the productive capacity of the planet Earth--if it has not already done so. The United Nations currently projects that--if strenuous efforts to implement family planning throughout the world are sustained for a century--that the global population may finally stabilize at a level of about 14 billion people in the 2090s. Any failure to address the problems of population limitation continuously during this very long period of time would make the final total still higher, assuming that the human race were not overtaken be extensive natural disasters first. The very large numbers of pre-reproductive individuals in developing countries indicated that their populations will continue to grow even if these individuals choose to have small families.

Poverty and malnutrition are so extensive in the world that it is difficult to imagine the orderly management of the resources that we have available. The World Bank estimates that a billion people--a fifth of the world's population--live in absolute poverty, defined as a condition of human existence below the lowest acceptable standards; and the United Nations calculates that some 500 million people, a tenth of the world's population, receive less than 80% of the recommended minimum number of calories--a diet inadequate for the proper development of the human body and brain. Developing countries, which comprise three-quarters of the world's population and must absorb nearly 90 million additional people with every passing year, control less than 15% of the world's goods, and must use their scarce earnings to try to manage their resources sustainably. The industrialized nations of the world, with a rapidly shrinking 25% of the world's population, control some 85% of the world's goods, and have shown only a limited understanding of our common problems, or willingness to act to ameliorate them. Thus, foreign development assistance, currently amounting to about 2% of the federal budget, is the least popular part of the budget for the great majority of Americans, even though we provide the lowest amount, on a per capita basis, of any industrialized nation, and have the most to lose through global catastrophe.

In visiting the countries of Central and South America, it is important to remember that they are experiencing a depression fully comparable, in the loss of human expectations, to that which affected the world in the 1930s. Only a few years ago, a great deal of economic advance seemed imminent, and the people of

Latin America looked confidently to their future. During the course of the 1980s, their dreams were largely shattered, their standard of living actually deteriorated, and their hopes of a better life for their children were substantially curtailed. Consider only one factor, the impact of the $1.2 trillion international debt. Brazil, for example, has repaid over $50 billion on the debt over the past ten years, consuming approximately 40% of its entire export earnings in the process, but not reduced the principal by a single cent! Against such a background, it is easy to understand the difficulties that the government of Brazil faces in addressing such fundamental problems as sustainable resource management, poverty, and the provision of basic services for a population of some 148 million people, growing currently by 3 million per year.

Tropical rain forest, which has now been reduced to about half of its former extent, occupies an area about three-quarters of the size of the contiguous 48 states of the United States. From this forest, an area about the size of Kansas is lost every year, either by clear-cutting or by destruction to the point where the alteration is irreversible. As a result of the combined effects of increased numbers of people, extensive poverty, and a collective unwillingness to address the problem, most of the remaining forest will have been reduced to small patches by early in the next century. More extensive areas will remain at that time, in areas such as the northwestern Brazilian Amazon and the Zaire Basin of Africa, but they too will be decimated by the middle of the 21st century unless there is a major change in attitudes worldwide. For these reasons, opportunities to visit and to conduct research in tropical countries should be seized as they present themselves, because they will never be as extensive in the future as they are now. The results of research can likewise be of direct importance in making possible the stable use of the forested lands, an outcome of great importance for humans.

Although tropical rain forests occupied only about 7% of the world's surface at their maximum extent, they probably contain a clear majority of the world's biodiversity. They are not the most threatened of tropical forests, actually, although their loss is the most highly publicized; seasonally dry forests, which usually grow on more fertile soil than rain forests, have mostly been cleared much earlier, to make way for cattle pastures and other forms of agriculture. At any event, only about 400,000 species of tropical organisms, including most of the plants and vertebrates-

-amounting to about half of the total--have been catalogued by scientists and named. Since efforts to sample the remaining diversity have yielded estimates ranging from 10 million to 80 million species, we may truly say that we do not even know the diversity of life on Earth--diversity that we are rapidly destroying--even to an order of magnitude. With the clearing of so many of the world's tropical forests over the next few decades, however, we may certainly say that we are likely to lose at least a quarter of the total number of species of plants, animals, fungi, and microorganisms that are living now during the lifetimes of a majority of us who are in a position to do something about it.

The loss of all these species--certainly well over a million--will be incalculably evil. Not only will they never be understood scientifically, but the aesthetic and cultural values that they possess will also be gone forever. Because we do not care enough, we are engaged in a massive denial of the Biblical injunction to be stewards of the world's wealth. The same species that we are losing are the elements that make the Earth's productivity sustainable: they make up the ecosystems that function over the entire surface of the globe, capturing a portion of the energy from the Sun and changing it into a form in which it can be used by organisms of all kinds, including humans. It is not a coincidence that tropical forests are so rich in species--a single square mile of Amazonian forest may be home to more than twice as many butterfly species as the estimated 700 that occur in the United States and Canada, and the West Virginia-sized country of Costa Rica has more bird species that that whole vast area--but rather a reflection of the fact that certain kinds of ecosystems function in specific circumstances, while others do not. The tragic consequences of our human failure to appreciate this relationship can be seen throughout the tropics in waste, degraded pastures; vast cut-over areas; and thousands of starving people, often moving from place to place in large numbers, the victims of ecological disaster.

Not only do organisms function collectively to regulate the flow of energy, originally derived from the Sun, and the cycling of minerals, but those same organisms are also the basis of all forms of productivity that ultimately support human populations. We draw our nutrition directly or indirectly from the photosynthetic processes that occur in plants, algae, and a few kinds of bacteria; and we obtain most of our food from a few dozen kinds of plants, despite the fact that tens of thousands of additional species are also capable of feeding us if we understand their properties sufficiently. Thousands of kinds of antibiotics have been patented since World War II alone, and chemicals derived from plants have become the basis for many of our most important medicines. In an age of genetic engineering, the loss of an individual kind of plant, animal, or microorganism represents not only the loss of a single species that might be useful to us, but the loss of a whole collection of tens of thousands of individual genes that could, in principle, be placed in other organisms and improve their useful properties.

For all of these reasons, our allowing so many kinds of organisms to become extinct can only be characterized as collective madness. We humans are consuming directly, diverting, or wasting some 40% of the world's total photosynthetic productivity at present, and have no strategy for dealing with the future, when our numbers double and then triple. Since one out of five people is living in extreme poverty now, and one out of ten is actually malnourished, our need to look carefully to the future should be evident: but we are neglecting to do so. What would the elements of a successful strategy to preserve as much as possible of the world's biodiversity be?

First, a series of parks and other protected areas should be created throughout the world's diverse regions, and funded internationally. Brazil and other nations of the developing world will not be able to spend their resources for such areas unless we help them by reversing the major cash flow from poor countries to rich ones. Secondly, seeds and other portions of plants and animals ought to be gathered and preserved outside of the communities in which they occur. In doing so, because of the incredible richness of life in the tropics, we must make careful selections, and save what we can because we have reasons for doing so. Seed banks and similar facilities can only accomplish so much; protected areas afford a much more practical and comprehensive strategy for preserving biodiversity. For either of these efforts to succeed, we must create an international system of laws and treaties that recognizes the central role of biodiversity in human affairs, and explicitly shares the responsibility for protecting it. Those who live in industrialized nations have a special responsibility to understand the problems and opportunities, the quality of life, of their fellow humans who live in developing countries. Visiting these countries and studying the magnificent biological richness they contain--a pleasing prospect that ought to be aided substantially by this remarkable book--is the one practical way of gaining the understanding that can motivate us to address these problems seriously. Above all, do not miss the opportunity to experience this richness while there is still time: we have the privilege of living in an age when more biological and cultural diversity exists than our children and grandchildren will have the chance to enjoy, to study, and to appreciate. Use the keys provided by this volume to take advantage of this irreplaceable opportunity.

Dr. Peter H. Raven

Missouri Botanical Garden
P.O. Box 299
St. Louis, MO   63166   USA

PREFACE

I have always been interested in natural history, especially insects. This interest, and National Geographic's alluring descriptions of scientific expeditions and their photographs of large and exotic insects, established early in me a desire to visit the tropics. This wish was first fulfilled in 1977 when I was awarded a scholarship by Earthwatch to participate on a study of social insects in Guanacaste, Costa Rica, led by Adrian Forsyth. It wasn't until 1983 that I was able to return to the tropics, again as an Earthwatch grant recipient. This time I assisted on a study of forest canopy insects, led by Terry Erwin at the Tambopata Wildlife Reserve in southeastern Peru. Participation on this project canalized my interest in tropical biology and opened my eyes to the necessity for rainforest conservation.

Since that time, I have visited the majority of countries in Central and South America to study insects and photograph nature. My academic colleagues and Latin American contacts have provided me with a great deal of information and advice regarding various tropical sites that can be visited, and where biological research can be carried out. However, this knowledge was obtained piecemeal over the years as a result of many correspondences, much investigation, and extensive traveling. In 1983, when I decided that my professional work as a biologist would be focused on the Neotropics, there was no reference book or guide available that presented a description of known sites and facilities. Today, seven years later, I still do not know of any such work.

I have written this book to fill that vacant literary 'niche'. Upon receipt of my Ph.D. in 1988, I was temporarily freed of academic and research obligations. Within a month, I embarked on a journey through the New World tropics with the specific purpose of visiting and evaluating rainforest sites and their facilities. My data were collected and recorded not only for the purpose of writing this book, but also to enable myself to judge and select those areas most beneficial for my own future research. I visited both tourist lodges and biological stations, which provided a wide degree of variation in comfort and services. I tried to select locations that were relatively accessible and safe, but the political climate of Latin America is such that conditions may change rapidly.

This book was written for anyone who has the desire to visit a rainforest for whatever reason. I have tried to describe accurately a variety of diverse tropical forest sites located throughout the Neotropics so that the reader (be he/she student, tourist, or naturalist) can make an informed decision as to which locality will best suit his/her purposes. Having a cautious and conservative nature, and remembering fully how it feels not to be able to speak the language of the people around you, it is my hope that the information contained within will alleviate some fears and apprehension about traveling in Latin America.

I have assumed that the reader already has a strong desire to visit a rainforest. For this book functions more as a guide and reference, rather than a descriptive piece or collection of essays. However, one of the most useful features presented within is the bibliography of various literature sources and organizations that can provide information about rainforests. The book section alone contains more than 200 references related to tropical forests, with titles of interest to both layman and biologist. For those wishing to get an intimate look at a rainforest community, there is a chapter describing 'hands-on' organizations. And finally, for students at the beginning of their career, there is a chapter on sources of funding for conducting research in the field of tropical biology.

A major factor that I have observed in my travels is how little effort is made by most tourist lodges with respect to providing conservation-oriented or educational exhibits and materials at their facilities. This seems incredible in light of the fact that their very business depends on the well-being of the rainforest and its flora and fauna. For this reason, I have formulated a set of recommendations for lodge owners. These recommendations will be sent to tourist lodge owners and managers, as well as being published in an appropriate journal article. I feel that nature-oriented tourism may be the most economically justifiable reason for the governments of developing countries to preserve their forests. I would like to participate in future efforts to establish ecologically sound nature tourism programs, with a strong focus on education and conservation.

Unfortunately, I feel that the tropical rainforests of the world may be doomed. Their rate of destruction is in the vicinity of 50-100 acres per minute. Whatever the actual number, it is a fact that such destruction will eliminate much of our planet's biological diversity, as well as cause major changes in global weather patterns. The next decade may be the last where sizable tracts of undisturbed tropical forests still exist. Only a large-scale involvement by the general public will have enough of an effect on law-makers, international banks, and governments to influence them into making more ecologically sound decisions. Although great strides have been made in the past few years resulting in a generally well-informed public, it takes more than a documentary or magazine article to motivate most people. I think an actual visit to a rainforest can provide the personal incentive that is needed. In that sense, I hope that this book will be responsible in aiding many laymen to actually visit and experience the beauty and wonder of a rainforest first-hand. With a large enough cadre of 'rainforest alumni', we may be able to convince the policy and decision-makers to act in a more responsible manner with respect to tropical forests, indigenous cultures, and future generations.

James L. Castner

INTRODUCTION

This book is divided into four chapters, followed by several appendices. Chapter 1 provides specific information about selected rainforest sites and their facilities in seven countries of South and Central America. The countries described are Peru, Ecuador, French Guiana, Venezuela, Trinidad, Costa Rica, and Panama. These were chosen because they contain tropical forests with tourist lodges or biological stations that are fairly easily accessible. Another important factor is that I have had the opportunity to visit and evaluate personally the sites I write about. For this same reason, I have not included Brazil, which contains the largest continuous tract of neotropical rainforest, nor Mexico, where the tropical forests closest to the United States exist.

Each country description begins with a map showing the capital and major cities, as well as the approximate location of the facilities and sites to be discussed. On the opposite page statistical data on the country's area, population, and capital is available, along with information about the language, currency, airlines, and telephone codes. There follows a brief, one page description of the country which provides general information including geography. Included in this is a list of the facilities that have been visited and evaluated.

Next follows the evaluation and description of specific sites. I tried to conduct my evaluations in a logical manner, recording that information which would be important to any tourist or biologist planning a visit. These data are organized and presented for each facility in the following order.

Contact

The name given may be a person or organization that is either the owner, manager, or administrator. It will however, direct the reader who to contact for additional information.

Address

This will be the mailing address of the owner/manager/administrator.

Description

This defines whether the facility is primarily used for tourist or research purposes.

Location

This gives the geographic location of the site within the country. Whenever possible, its distance from nearby large cities has been included, as well as its elevation and approximate latitude.

Logistics

In this section I have tried to give a fairly detailed account of how one reaches the facility under discussion. Many times there are several methods available, but my discussion is usually limited to the mode of transport that I used personally. The starting point used is often the country's capital, or the nearest city to the site that maintains a commercial airport. I have included the time durations for various portions of the journey based on personal experience.

Forest Type

I have followed the Holdridge Life Zone Classification whenever possible here. This classification system defines and names recognizable areas of vegetation based on rainfall, temperature, latitude, and altitude.

Seasonality

This gives a breakdown of the 'wet' and 'dry' seasons, the months in which they occur, and the usual amount of annual rainfall.

Facilities

Included here is the construction, furnishing, and layout of dining, sleeping, bathing, and toilet facilities. The presence and type of electricity is also reported. In the case of biological stations, specialized scientific equipment may be described.

Trail System

Under this heading I have described the presence and extent of any organized system of trails. Also provided is information on whether and how the trails are marked, and if trail maps are available.

Costs

I have tried to provide figures that were accurate at the time of my stay at these various facilities. I've also indicated that some lodges are willing to extend a special price to biologists, especially for a long-term stay. However, everyone should verify their rates personally, before their visit!

Comments

In this section I describe things or events that made a particular impression on me. I emphasize that this is solely my opinion, based on my experiences.

Interspersed among the descriptions are photographs that I have taken of the facilities. Following the evaluations of sites I have visited in a particular country, may be information about other facilities that I have not seen personally. This information was derived from correspondences, brochures, and discussions with colleagues. Therefore, the amount of data available will vary greatly, and sometimes be limited to only an address.

Next, I have usually provided a brief description of the country's capital city, and in some cases other major cities that one must pass through in order to reach the areas described. This section will include comments on the airports, hotels, and transportation. The purpose is to prepare the traveler for what to expect. Information is not presented like a tourist guidebook, and hotels are mentioned only to help out first-time travelers to an area that might not have any idea of where to go.

The remaining sections under each country's description give information on books, maps, tourist information sources, conservation organizations, and scientific organizations/institutions. Under books, I have provided a list of references that specifically treat the country in question. Some are listed again later in Chapter 2 as part of the 'rainforest bibliography', while others are not. The map section tells where country, road, and sometimes national park maps can be purchased. Usually this is at the geographic institute. The remaining sections present names and addresses where the reader can secure additional information.

Chapter 2 deals with rainforest information sources. A partially annotated rainforest bibliography composes the main body of this chapter. The books listed are grouped into the following eight categories: 1) Natural History And Ecological Books For Laymen, 2) Adventure, Travel, And Exploration Books (1950-present), 3) Adventure, Travel, And Exploration Books (pre-1950), 4) Books For Biologists, 5) Flora And Fauna Guides, 6) Environmental, Political, And Miscellaneous Titles, 7) Travel Guides And Regional Titles, and 8) Statistical And Reference Books. This is followed by a section: Magazines For Naturalists and Journals For Biologists. Next, a short listing of suppliers of high quality maps of Latin America is provided.

The last portion of Chapter 2 lists and describes organizations that are primarily conservation-oriented and involved in rainforest-related work. They may actively support ecological research by their staff biologists, or they may be devoted to public awareness goals and work mainly with the media. All, however, employ personnel with interests and experience in various aspects of the biology, ecology, and conservation of rainforests.

The serious naturalist may wish to examine more closely the rainforest community, and sometimes to take part in actual scientific studies of it. Many people today want to do more than 'spend a night in the jungle' and then move on to the next excursion. The beginning student of tropical biology may wish to gain 'field experience' and see some of the biological interactions and mechanisms that they have read about. Laymen in general may wish an educational and exciting alternative to the average 'fun in the sun' or 'sightseeing' vacation. Chapter 3 discusses how this can be accomplished with a review of 'hands-on' organizations. These organizations frequently offer opportunities to visit and 'work' in tropical forests through courses, study programs, and volunteer-staffed and -funded scientific research programs.

Finally, Chapter 4 is titled Sources Of Funding and is aimed primarily at the academic community. It was written in the hopes of providing beginning graduate or undergraduate students with a list of organizations that could potentially fund their original research on some aspect of tropical biology. Based on personal experiences and those of colleagues, I have included the major sources that fund biological research.

The appendices supply further information on subjects including traveling, specialized vocabulary, tropical biologists, and zoological and botanical gardens. The first appendix offers tips to make traveling in Latin America less strenuous and more enjoyable. It also provides a list of selected travel agencies that offer or specialize in natural history tours. The second gives a list of words in English and Spanish that may be of particular use to biologists working on a rainforest related project. The appendix on tropical biologists lists scientists according to discipline. It provides names, affiliations, and mailing addresses. This will permit correspondence and give the person interested in securing technical literature on a subject a few names to start with in the Science Citation Index, or other sources.

No section of this book is meant to provide total or definitive information on any given subject. Anyone who has written a term paper or researched a topic in the library knows that one reference leads to another. Soon this has a snowballing effect, but it is often difficult to find those initial references. Hopefully this book has done an adequate job of providing a starting point for information on a variety of subjects. Its main purpose is to make it easier for anyone to visit or study the rainforests of Central and South America in any capacity.

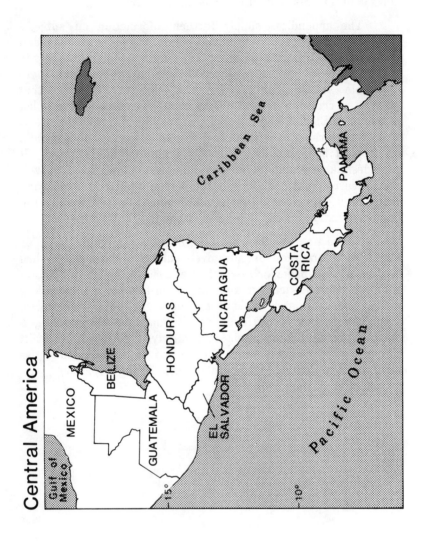

South America

RAINFORESTS

COUNTRIES, SITES, AND FACILITIES

CHAPTER 1A   PERU

General Information

Area: 1,285,200 sq km (496,000 sq miles)

Population: 21,790,000

Capital: Lima (population: 5,000,000)

Language: Spanish

Currency: Inti (1 Inti = 100 céntimos)
            Coins: 1, 5, 10, 25, and 50 céntimos
                   1 and 5 intis
            Notes: 10, 50, 100, and 500 intis
            Exchange Rate: U.S. $1 = 2,000 Intis (January 1989).
                           U.S. $1 = 12,800 Intis (March 1990).
            (Until 1986 the official currency in Peru used to be the sol. Although intis have almost completely replaced soles, there are still some notes in soles in circulation. One inti is worth 1,000 soles.)

National
Airlines: AeroPeru, Faucett

Telephone   Peru   51
Codes: Lima   14
            Cuzco   84
            Iquitos 94

Peru

1 – Manu Lodge 2 – Explorers Inn 3 – Cuzco Amazonico Lodge
4 – Explorama Inn 5 – Explorama Lodge 6 – Explornapo Camp

Introduction

Peru is the third largest country in South America and is located in the northwestern portion of the continent. It is bordered on the north by Ecuador and Colombia, on the east by Brazil and Bolivia, on the south by Chile, and on the west by the Pacific Ocean.

Peru has three distinct geographical regions. Along the western coast is a narrow desert consisting of arid plains and foothills. It takes up only about one tenth of the country's area, but supports nearly half of the population. The remaining portion of the western half of the country is composed of the Andes Mountains, which generally parallel the coast. These mountains cover approximately one fourth of the country, but are inhabited by half of the total population. The remaining region is the vast *selva*, including the Amazon Basin. It is an area of tropical forests and rivers that covers almost two thirds of the total land area and the entire eastern half of Peru. However, only a small portion of the total population lives in this region, usually in settlements along or near the rivers.

The city of Iquitos, located in the northeastern department of Loreto, is Peru's most famous port on the Amazon River. Although 3,200km from the mouth of the Amazon in Brazil, the river at Iquitos is still navigable by large vessels. The city of Cuzco, located in the south-central department of the same name, is visited by thousands of tourists each year on their way to the Inca ruins at Machu Picchu.

My evaluation of rainforest sites in Peru is based primarily on information obtained and observations made during January-February 1989. At this time I visited three facilities in the lowlands of southeastern Peru (Manu Lodge, Explorer's Inn, and Cuzco Amazonico), and three facilities in northeastern Peru (Explorama Lodge, Explorama Inn, and the Explornapo Camp). Brief information about other lodges near these locations is also provided.

Manu Lodge

Contact

    Boris Gomez

Address

    Manu Nature Tours
    Avenida Sol 627-B
    Oficina 401 and 202
    Cuzco, Peru
    Telephone: 23-47-93
    Telex: 52003 or 52004
    Fax: 084-23-47-93

Description

    Manu Lodge is a tourist lodge that was constructed by Manu Nature Tours in 1988.

Location

    The Manu Lodge is located on the edge of an oxbow lake(Cocha Juarez) formed by the Manu River. It is situated in the Department of Madre de Dios in southeastern Peru, approximately 125 km northeast of the city of Cuzco, and approximately 135 km west/northwest of Puerto Maldonado. Manu Lodge is situated between 12°-13° South latitude at an elevation of approximately 400 meters.

Logistics

    Manu Nature Tours offers two alternatives for reaching the Manu Lodge, both of which originate in Cuzco. The quickest route is by plane. The aircraft used during my visit was a twin-engine Cessna 402 capable of taking 8 passengers. (A larger plane capable of accomodating 13 passengers was expected to be in service in the near future.) The plane flies from Cuzco over the Andes to a narrow gravel landing strip in the jungle near the village of Boca Manu, situated a short distance from the confluence of the Manu River with the Alto Madre de Dios River. The flight takes approximately 45 minutes and is followed by a 10 minute walk to where the boat is moored on the bank of the Alto Madre de Dios River. People were available at the landing strip to carry luggage and equipment. (Note: There may be a weight limitation to the amount of luggage one can bring on this flight.)

An 18-meter tarpaulin-covered, outboard motor-powered boat was used to transport both passengers and the supplies and luggage. The boat initially travels up the Alto Madre de Dios River and quickly passes the village of Boca Manu. At this point the boat heads up the Manu River towards the lodge, a trip of $3\,1/2$ - $4\,1/2$ hours depending on the condition of the river. After about one hour, the park guard station called Romero is reached. Here passengers must disembark and sign the register book for visitors. Previously, a $10 visitor fee was paid at this station during the return trip when visitors left, but now this is paid in Cuzco. After continuing the boat ride for another $2\,1/2$ - $3\,1/2$ hours, a destination on the east bank of the Manu River is reached. From here visitors walk approximately 20 minutes along a forest path to the edge of Cocha Juarez. From the dock at this point a catamaran-like boat, fashioned from two dugout canoes and a platform is paddled for the 15 minute trip to Manu Lodge.

The alternative route to Manu Lodge includes both overland and river travel. Manu Nature Tours uses their own new 27-seat luxury bus to transport visitors from Cuzco, over the Andes and to the Alto Madre de Dios River. This overland-river route takes the better part of two days each way, but provides the additional perspective of entering the Manu Biosphere Reserve by descending from the mountains to the lowlands. This allows one to see the spectacular timberline cloud forest of Tres Cruces at the upper limit of Manu National Park. Visitors taking the overland route usually overnight at the end of the first day of travel at a lodge located near the village of Atalaya.

Forest Type

The area adjacent to Manu Lodge is classified as Tropical Moist Forest. However, the Manu Biosphere Reserve encompasses a vast area with elevations reaching 3,600 meters above sea level providing a variety of forest habitats. The forests on alluvial soil near the Lodge are quite different from those on lateritic soil.

There is also a large array of different natural successional stages along the river.

Seasonality

The average annual rainfall at Manu is approximately 2,000mm. The wet season extends from about November to March. The dry season occurs from approximately April through October, with the driest month normally August.

Facilities

Manu Lodge consisted of several elevated buildings constructed primarily of mahogany (*Cedrela odorata*) logs taken floating in the Manu River. The main lodge building had a thatched roof and nylon screening to exclude insects. A long wing of this building consisted of 10 guest rooms, each with two single beds equipped with mosquito netting. Every guest room also contained a desk and chair, shelves, small tables, and hooks and

The Manu Lodge located in the Manu Biosphere Reserve. (In the Department of Madre de Dios, Peru)

hangers. All boots and shoes were left in two screened porches located at the entrances. The other portion of the lodge building consisted of a dining room and bar, and an adjacent area where books, magazines, and a short-wave radio were kept. Two successively higher levels were located above the bar area and were capable of comfortably seating 6-8 visitors. These upper levels provided a view of the lake and trees in front of the lodge where there was often much bird activity. A separate kitchen was located behind the dining area and was connected to the main lodge building by a covered, screened walkway.

Bathing facilities consisted of a separate raised building located near the wing of the lodge containing the visitors' rooms. There were six shower stalls in a covered, screened building within 15 meters of the lodge. Each shower room also contained a sink. Water to the shower building was pumped up from the lake. Towels and soap were provided.

Toilet facilities were located approximately 25 meters from the lodge, just inside the forest. These consisted of three latrines located in separate large screened buildings with roofs. A raised box with a toilet seat fastened to it, located over a pit formed the toilet in each.

During my visit there was no electricity at Manu Lodge. Lighting in the buildings after dark was accomplished by candles and kerosene lanterns. A small portable electric generator was also present for running some appliances.

Food at Manu Lodge was simple, but good. Boxed lunches or snacks were available to visitors going out on long hikes or day trips. Containers of filtered water were available near the guest rooms and the dining area. Thermos bottles filled with hot water for tea and coffee and plastic pitchers filled with powdered orange drink were generally available throughout the day.

Trail System

A number of distinct trails led out from Manu Lodge and through the surrounding area. There were approximately 7km of trails, with plans to extend the network to about 20km. Trails were marked with somewhat faded red and orange flagging tape. Simple trail maps are presented on some of the brochures put out by Manu Nature Tours. Guides were provided to visitors at no extra charge.

Costs

A number of tour plans exist for visiting Manu Lodge, including different methods of transportation and lengths of stay. The cost will be dependent upon these previous factors, as well as the number of people traveling in your group. In general however, it may be slightly more expensive to visit Manu than other lodges and areas in Peru, due to its inaccessibility. My 4 day/3 night tour cost $403 during January 1989, which included transfers from the airport and round trip airfare from Cuzco to Boca Manu.

Biologists and researchers planning long-term studies should discuss their intended work with Boris Gomez in advance. A discount may be offered to legitimate researchers wishing to stay at Manu Lodge for extended periods.

Comments

Visitors to the Manu Lodge probably have a better chance of seeing monkeys, large mammals, and birds there, than anywhere else in the western Amazon region. During the 3 1/2 hour boat trip from the Boca Manu air strip to Manu Lodge, our group sighted 157 macaws of four different species. During my 4 day/3 night visit, I also saw four species of monkeys including the emperor tamarin. Other rare animals such as the giant river otter, can commonly be seen within a short distance of the lodge.

The lodge itself is solidly constructed and designed. It was not made to appear rustic, yet the thatched roof allows it to blend in with the surrounding forest. It is open and airy within and does not give a claustrophobic feeling that I have found in so many lodges that try to imitate native architecture. The practice of leaving one's footwear on the porches allows the inner floor to remain very clean so that visitors can comfortably walk about without shoes inside.

The guide during my visit was a Peruvian whose main expertise was in spotting and identifying birds and monkeys. His practice of carrying a tripod-mounted spotting scope, allowed for exceptional views of difficult to see birds and primates. Researchers assisting in scientific investigations near the lodge sometimes also act as guides. Boris Gomez, his guides, and his office staff were all bilingual.

The Manu Lodge was constructed to accomodate only 25 visitors in order to prevent any damaging impact to the surrounding forest. I received the distinct impression that Mr. Gomez is sincerely concerned with conservation and that he wishes his operation to benefit both tropical biology research and rainforest preservation. Manu Nature Tours maintains close ties with the Conservation Association for the Southern Rainforest of Peru (which has its main office in the same building in Cuzco as Manu Nature Tours). In addition, logistical support has been provided at Manu Lodge to scientists who work both from the lodge itself, and from the biological stations (such as Cocha Cashu) located further upriver. This results in the Manu Lodge staff remaining constantly informed about the movement of local animal populations which are of interest to visitors.

The inacessibility of Manu Lodge, combined with the protected status of the Manu Biosphere Reserve and the low population of the surrounding region, have resulted in this particular area of Peru remaining virtually pristine. This same inacessibility is responsible for greater logistical difficulties and slightly higher costs necessary to reach the lodge.

Visitors interested in camping at Manu may make arrangements through Manu Nature Tours or through the tour operator listed below:

Expediciones Manu
Procuradores 372
Cuzco, Peru

Explorer's Inn

Contact

Dr. Max Gunther

Address

Peruvian Safaris S.A. Garcilazo de la Vega 1334 P.O. Box 10088 Lima (1) Peru OR Telephone: 31-63-30 Telex: 20416 PE SAFARI Cable: SAFARI	Peruvian Safaris S.A. Plaza San Francisco Segundo Piso Cusco, Peru Telephone: 23-53-42

Description

The Explorer's Inn is a tourist lodge whose construction was started in 1976 by Peruvian Safaris.

Location

The Explorer's Inn is located along the Tambopata River near the confluence with the La Torre River, in the department of Madre de Dios in southeastern Peru. The Explorer's Inn is approximately 27km (straight-line distance) west/southwest from the departmental capital of Puerto Maldonado. The distance by river from Puerto Maldonado to the Explorer's Inn is approximately 57km. The Inn is situated at approximately 260 meters elevation at about 13° South latitude.

Logistics

To reach the Explorer's Inn, one usually begins by traveling to the jungle town of Puerto Maldonado. Air service to Puerto Maldonado from Lima via Cuzco is provided by both of Peru's major airlines: Faucett and AeroPeru. (Airfare from Lima to Puerto Maldonado can be quite inexpensive when the same carrier is used internationally to reach Lima.) There is also access to Puerto Maldonado by road from Cuzco, however this method is not recommended due to unreliability of drivers and unsafe road conditions.

Guests are met at the airport in Puerto Maldonado by staff members of the Explorer's Inn. Transfer to the boat dock takes approximately 15 minutes via private bus or vehicle. Visitors then embark on a ride of approximately three hours up the Tambopata River, in a covered, outboard motor-powered boat.

Forest Type

The Explorer's Inn exists in a zone between Tropical Moist Forest and Sub Tropical Moist Forest. There is a variety of floristically very different vegetation types including swamp forest, bamboo forest, and forests on alluvial, clay, and sandy clay soils.

Seasonality

Annual rainfall at the Explorer's Inn usually ranges from 2,600mm-3,200mm. The rainy season normally starts in October and extends through to April, with January and February the wettest months. The dry season begins in May and progresses through to September, with July and August normally the driest months.

Facilities

The Explorer's Inn and its facilities exist in a clearing of 2-3 hectares at the edge of the forest, about 100 meters in from the Tambopata River. These facilities are located within the Tambopata Wildlife Reserve, a 5,500ha ecological reserve with the status of 'special protected area' that has been set aside by the Peruvian government. Buildings were located around the edge of the clearing with the lodge in the center. This main lodge building was a large, raised, roughly octagonal structure constructed of natural materials and resembling a large native hut. It had a thatched roof and a covered veranda with tables and chairs around several sides of its perimeter. Inside were a small bar and numerous rough-hewn dining tables and chairs.

At one side of the lodge, accessible to visitors, was a library featuring many books on flora and fauna, tropical biology, and conservation. Files of photographs of local plants and animals taken by scientists who have worked at the Explorer's Inn were also available. On display were examples of tropical fruits found on the Tambopata Wildlife Reserve, as well as a poster with excellent drawings of indigenous animals.

A spiral staircase radiating about a large tree trunk led to the second floor of the lodge. This upper level was used to present a conservation message and had on display a number of informative posters discussing rainforests. Brochures and circulars discussing rainforest conservation groups were available.

All meals were taken in the lodge, unless a boxed lunch or breakfast was specifically requested. Meals were served to visitors at their tables unless the Inn was filled near capacity, in which case meals were buffet style. Food was simple and adequate. Meal times were approximately 5:30 am, 1:00 pm, and 7:30 pm. The bar offered soda, beer, and liquors for sale. There was usually hot water available in thermos jugs for tea and coffee 24 hours a day. Potable boiled water was also always available in the main lodge building.

The Explorer's Inn was capable of lodging a total of 74

The Explorer's Inn lodge building at the Tambopata Wildlife Reserve. (Near Puerto Maldonado, Peru)

A guest bungalow at the Explorer's Inn. (Near Puerto Maldonado, Peru)

people. Rooms for tourists at the Explorer's Inn were situated in seven bungalow buildings located on either side of the lodge. Bungalows were elevated approximately a meter off the ground, had thatched roofs, and a wide covered veranda in front with chairs and tables. Most contained four rooms, two with two single beds and two with three single beds. Rooms were screened and no mosquito nets were provided (nor did biting insects in the room appear to be a problem). Rooms also contained a chair, end table (with candle, holder, and matches), and shelves for clothing and belongings.

A bathroom with sink and shower were also present in each room. Both sink and shower used filtered water, originally pumped up from the river. Visitors were told that this water is acceptable for brushing one's teeth, but that it should not be swallowed or used for drinking water. The water was turned on early in the morning and turned off at approximately 9:00-10:00 pm at night.

There was no electricity at the Explorer's Inn. Several kerosene-powered refrigerators existed for food storage and keeping refreshments cold.

A large plastic tub and soap powder could be obtained from the management for visitors who wished to hand wash their clothes. The Explorer's Inn also provided rubber boots to guests for use when the trails were muddy.

Trail System

There was an excellent network of approximately 30km of trails on the Tambopata Wildlife Reserve that led throughout a variety of habitats. The trails were marked (originally every 50 meters) by 10cm diameter white disks attached to trees at approximately breast height. These disks bear an abbreviation of the trail name and the distance in meters from the trail's beginning. Trails were well defined and easy to follow. Visitors were provided with a detailed trail map on request.

Resident Naturalist Program

Peruvian Safaris attempts to employ a minimum of four resident naturalists at the Explorer's Inn at all times. These naturalists are usually of European or U.S. origin and generally have a background or undergraduate degree in some aspect of biology. Resident naturalists are usually contracted for a minimum stay of three months at the Explorer's Inn, although many stay longer and are encouraged by the management to do so. Most naturalists conduct a research project during their residency at the Tambopata Wildlife Reserve.

The resident naturalists take turns acting as guides for tourist groups visiting the Explorer's Inn. During several planned excursions, they point out and discuss various general characteristics of the rainforest community, as well as examples of the local flora and fauna that are of particular interest. Attempts are made to match specific guides with appropriate special interest groups. There is nearly always a resident naturalist with ornithological expertise present. Often some of the most highly respected bird authorities in the world conduct tours at the Tambopata Wildlife Reserve due to the record number of species that have been observed a short distance from the Explorer's Inn.

Costs

The cost of staying at the Explorer's Inn depends on the season (high season = April 15 - December 14, low season = December 15 - April 14), the length of stay, and the number of people in one's group. For example, based on a price list I obtained at the Explorer's Inn during January 1989, a 'jungle adventure' package of 3 days/2 nights for a group of 1-4 people cost $150 per person in the high season and $135 per person in the low season. This included land and river transportation from and to the Puerto Maldonado airport and room and board at the Explorer's Inn. A variety of plans and packages are available and prices and reservations should be confirmed with Peruvian Safaris.

Dr. Max Gunther and Peruvian Safaris have a history of aiding legitimate researchers by offering discounted rates and logistical aid. Scientists conducting approved research typically pay only $36 per night. Biologists wishing to conduct long-term investigations (> one month) at the Tambopata Wildlife Reserve from the Explorer's Inn may qualify for a greater discount and should discuss this option with Dr. Gunther.

Comments

World record numbers of birds and of several kinds of invertebrates including butterflies, have been identified from the Tambopata Wildlife Reserve. Some may argue that this is due to the intensive research that has been carried out there during the past decade, but the habitat diversity is also important. It cannot be denied that southeastern Peru in general is one of the richest and most biologically diverse areas in the world.

The operation of the Explorer's Inn by Peruvian Safaris appeared smooth and well run. Additional buildings have been constructed on a site behind the lodge and are used as an administrative office and workroom for the naturalists. Weather data and information and journals of a more scientific nature are kept in the naturalists' office.

One of the most appealing features of the Tambopata Wildlife Reserve is the well-marked and mapped trail system. Laminated copies of trail maps allow visitors to easily use the trail network with little fear of getting lost. People wishing to wander about on their own may do so, or follow a self-guided nature tour. The system of trail marking also allows visitors to consult records of naturalists about faunal sightings, and subsequently locate and visit the approximate locations of interest.

Cuzco Amazonico

Contact

    Jose Koechlin von Stein

Address

    Albergue Lodge Cuzco Amazonico
    Andalucia 174
    Lima 18, Peru
    Telephone: 46-27-75 or 46-97-77 or 47-71-93
    Telex: 21475 PE LATIN

Description

    Cuzco Amazonico is a combination tourist lodge and privately maintained ecological reserve.

Location

    Cuzco Amazonico is located on the east bank of the Madre de Dios River, approximately 15km east of Puerto Maldonado in the department of Madre de Dios in southeastern Peru. It is situated at an elevation of approximately 200 meters above sea level, between 12°-13° South latitude.

Logistics

    Cuzco Amazonico provides transportation for its guests from and to the airport in Puerto Maldonado. This city can be reached by daily jet service offered by either of the Peruvian airlines Faucett or AeroPeru. Cuzco Amazonico representatives meet incoming visitors and drive them to the boat dock on the Madre de Dios River approximately 15 minutes away in their own bus. Passengers and luggage then go on a covered, 10-meter, outboard motor-powered boat which takes them to the lodge. The trip downriver on the Madre de Dios from Puerto Maldonado to Cuzco Amazonico usually takes about 50 minutes, while the return trip upriver normally lasts about 1 1/2 hours.

One can also reach the city of Puerto Maldonado by road from Cuzco, but this route is not recommended due to the dangerous condition of the road and the difficulty in finding a reliable driver.

Forest Type

The forest at Cuzco Amazonico is classified as Sub Tropical Moist Forest. Around the Lodge it is mostly on alluvial soil and many of its species are shared with similar areas of Manu Park. There are also numerous patches of swamp forests with other vegetation types, easily accessible across the river.

Seasonality

The climate at Cuzco Amazonico is divided into distinct dry and rainy seasons with the former more pronounced than at Tambopata or Manu. The dry season occurs from about May through October. The wet season lasts from November through April. The annual rainfall is approximately 2,000mm.

Facilities

The Cuzco Amazonico lodge and its facilities were located in a clearing on the north/east bank of the Madre de Dios River. The main lodge building began within 50 meters of the boat dock and had a long, covered hallway on one side that led to the visitors' cabins. The lodge building served as the dining room, bar, and lounging area. It was constructed primarily of natural materials in the general style of native architecture. The building proper was raised off the ground. There was a covered, open-air porch with tables and chairs in the front, while the main dining area was screened and could provide seating for 80-100 people.

Several books were kept accessible to visitors on tables at one side of the dining room. These included the leading guide books to the birds of South America and a photographic compendium of the reptiles and amphibians found at the reserve.

All meals were taken in the dining hall. Dinner meals were good and usually consisted of native dishes, while breakfasts tended to be light. Meal times were approximately 5:30 am, 12:30 pm, and 7:00 pm. Visitors could obtain coffee, tea, or other refreshments throughout the day by asking one of the lodge employees. Soda, beer, and liquors could be purchased from the bar.

Cuzco Amazonico was capable of accommodating approximately 100 visitors. Lodging was provided in 50 elevated cabins located in a clearing a short distance from the lodge building. Sawn cross sections of tree trunks formed a walkway between cabins. Cabins were set up as doubles, triples, or for family groups. The features of my double cabin were as follows.

Several steps led up to a small, open-air, covered porch which supported two hammocks. The room itself had screened window areas on three sides, but open eaves which allowed easy access to insects. Construction was of natural appearing materials with a thatched roof. The main portion of the room interior was occuppied by two single beds with mosquito nets, and a large low

Guest cabins at the Cuzco Amazonico Lodge. (Near Puerto Maldonado, Peru)

table or luggage rack placed between them.

A carved wooden sink stood in one corner of the room. To the side of this sink, and separated from the sleeping area by a wall partition, was a flush toilet and shower stall. Water for the showers and sinks is pumped directly from the river and stored in a holding tank. Bottles of potable water are provided in each room for brushing one's teeth and drinking. This water originated in Puerto Maldonado and was treated with chlorine. The shower/sink water was turned on about 5:00-5:30 am and shut off in the evening around 7:00 pm.

There was no electricity at Cuzco Amazonico. Each day before twilight, an employee lit a kerosene lamp which was left on the cabin porch. Visitors could then suspend this lamp from an overhead beam inside the cabin. Kerosene pots between cabins were also lit at twilight.

A laboratory building and scientists' dormitory were constructed in 1989 for use by students and researchers visiting Cuzco Amazonico. The former is an elevated, screened structure approximately 6m x 14m, which contains generated light through fluorescent bulbs, counter space with benches, running water, and a bathroom.

Trail System

There were approximately 10km of trails at Cuzco Amazonico. These began on either side of the main lodge. Trails were normally marked approximately every 100 meters with 5cm metal disks nailed to trees. Trail maps were not handed out to visitors, but those interested in the trail layout and explanation of the numbering system can consult a spiral-bound booklet titled *Informe Preliminar de Actividades*. This booklet was kept in the lodge building with the bird guide books and offered a map of the trails on the reserve on page 109 (Figure 1).

Three Peruvian nationals were employed as guides during my visit, and were responsible for taking tour groups out on trails and explaining the forest. Rubber boots were sometimes available to tourists staying at the lodge, but one shouldn't plan on a pair that fits being available.

Costs

The costs of staying at Cuzco Amazonico vary with the length of stay and the number of people in one's group. For example, 1-4 persons may stay 2 days/1 night for $105 each, 3 days/2 nights at $143 each, or 4 days/3 nights for $200 each. This includes reception at the airport in Puerto Maldonado, transfer by bus or car to the boat dock and then by canoe to the Cuzco Amazonico Lodge, guided tours according to chosen programs, lodging in twin bedrooms with bath at the lodge, overnight in a large communal room when camping, and all meals (no drinks). A single supplement is an additional $26 per night.

Biologists and long- or short-term students whose research or studies have been approved and previously arranged by the Cuzco Amazonico staff are offered a discounted rate of $25 per night. This includes lodging, transfers, and full board.

Comments

The Albergue Lodge Cuzco Amazonico hosts tourists from all over the world, but with a high percentage coming from Europe. Excursions available to visitors included a combination boat trip and hike to nearby Lake Sandoval. During this trip hoatzins and other water fowl could be seen, as well as monkeys.

A number of different scientific projects have been conducted at Cuzco Amazonico. The reptiles and amphibians have been especially well worked on and a photographic/informational compendium of those collected on the reserve has been prepared by Dr. William Duellman.

The trail system was not very extensive and mostly restricted to the area near the lodge where numerous alternate routes can be confusing. Some of the numbered metal disks used as markers were absent, but most trails also had numbered flags. The 'long' trail which I walked was not very wide, and in places not very distinct. Four kilometers of trail pass through a series of 1ha plots in which all trees and lianas greater than 10cm diameter have been identified and mapped as part of the BIOTROP project of the University of Kansas and the Missouri Botanical Garden.

The **Tambo Lodge** is another tourist facility located in the Puerto Maldonado area. It is located on the banks of the Madre de Dios River, approximately 8km southeast of the city of Puerto Maldonado. For more detailed information try contacting the following addresses:

Tambo Lodge Travel Agency   Tambo Lodge Travel Agency
Portal de Panes 109   Hostal Moderno
Plaza de Armas   OR   Billingurst 359
Cusco, Peru   P.O. Box 146
Telephone: 23-61-59   Puerto Maldonado, Peru
                                    Telephone: 63

Explorama Lodge

Contact

    Peter Jenson

Address

    Explorama Tours
    Box 446
    Iquitos, Peru
    Telephone: 23-34-81 or 23-54-71
    Telex: 91014
    Fax: 51-94-23-49-68

U.S. Representatives:

    Selective Hotel Reservations
    Telephone: (800)-223-6764 (Nationwide)
        (617)-581-0844 (Inside Massachusetts State)

Description

The Explorama Lodge is one of three tourist facilities maintained by Explorama Tours. It was constructed in 1965 and many of its buildings have been completely renovated during 1987-1989.

Location

The Explorama Lodge is located approximately 80km down the Amazon River in a generally northeasterly direction from the city of Iquitos, the capital of the department of Loreto in northeastern Peru. The Lodge is situated at an elevation of approximately 130 meters and is located between 3°-4° South latitude.

Logistics

To visit the Explorama Lodge one must first get to the city of Iquitos. Both the Peruvian airlines Faucett and AeroPeru have regularly scheduled flights between Lima and Iquitos. Faucett has also maintained one flight per week that flies directly from Miami to Iquitos, and Iquitos to Miami.

Arriving guests were met by representatives of Explorama Tours at the Iquitos airport. Transportation was then provided by private bus for the 15-20 minute ride from the airport to the boat dock on the Amazon River. Located directly below the administrative offices of Explorama Tours, one of their fleet of boats took guests downriver to the Explorama Lodge. The most common type of boat used was an outboard motor-powered, thatched-roofed vessel of about 10 meters length and capable of holding 25-30 people. The downriver boat trip from Iquitos to the Explorama Lodge took about 2 1/2 hours. The return trip, going against the current from the lodge to Iquitos, usually lasted approximately 3 1/2 hours.

During the rainy season, the boat journey terminated by traveling up a small tributary of the Amazon called the Quebrada Yanamono, where there was a floating landing platform. The lodge facilities were immediately adjacent to the platform. During the dry season this tributary may not be deep enough for the boat to use, necessitating a 15-20 minute walk for visitors from the Amazon River to the Explorama Lodge. Guests needed only carry personal items, as lodge employees handled suitcases and all heavier supplies.

Forest Type

The forest near the Explorama Lodge and surrounding areas is classified as lowland Tropical Moist Forest. Near the lodge are both seasonally inundated varzea forest and upland forest on clay soil. The latter has the most diverse tree flora in the world, with 300 species of woody plants greater than 10cm diameter out of the 600 individuals in a one hectare plot studied by Dr. Al Gentry of the Missouri Botanical Garden.

Seasonality

The area around the Explorama Lodge receives approximately 3,500mm of rain per year. There is no definite dry season, although the river is lowest from June to January and highest from February to May. Every month of the year (except May) has been the wettest, and most months have been the driest of at least one calendar year during the past three decades for which records are available. The Amazon River has its peak water levels in the Iquitos region usually during the months of April and May.

Facilities

The Explorama Lodge is located on the 195ha Bushmaster Reserve that is owned by Explorama Tours. However, Explorama Tours also effectively controls and manages the 1600ha Yagua Reserve nearby as well. Due to recent increases in the levels of flooding during the wet season, many of the facilities at the Explorama Lodge have been reconstructed on higher ground. The new dining hall was an elevated, screened building with a high thatched roof to dissipate the heat. Kitchen facilities adjoined the dining hall and a bungalow for lodge employees was planned for behind the kitchen. The dining hall was furnished with comfortable tables and chairs and can accocmodate 100 guests at one sitting. Adjacent to the dining hall (in the same building) was a separate bar/lounge area where sodas and beer could be purchased. Immediately outside the bar was an open-air, covered porch with tables, chairs, and benches. A wide covered veranda ran the length of the dining hall.

Food at the Explorama Lodge was excellent. Fresh fruits and fish were obtained locally from the Iquitos area, while fresh vegetables and other items were shipped in from Lima. The meal times were signalled by the beating of a large drum that could be heard throughout the lodge. There were always thermos jugs of hot water available, along with tea, instant coffee, and instant hot chocolate. Guests had access to these items 24 hours a day. A dispenser of purified water for drinking was also available at the dining hall (as well as at the bungalows).

A second bar (La Tahuampa) where soda, liquors, and beer could be purchased is located on the opposite side of Quebrada Yanamono from the dining hall. It also featured and sold native handicrafts, including Shipibo pottery and cloth paintings. Many lodge employees were excellent musicians and entertained guests with their music in the La Tahuampa bar from late afternoon until several hours past dinner.

The Explorama Lodge had the capacity to accomodate 144 people. Guests were housed in four bungalows which had a total of 72 rooms. The guest rooms generally contained two single beds with mosquito nets. There were also a set of shelves, wooden stools, and a table with a basin for washing one's face and hands. A fresh pitcher of 'washing' water obtained from the *quebrada* was placed daily in each room. The rear of these rooms

An elevated guest bungalow at the Explorama Lodge. (Near Iquitos, Peru)

was an unscreened window that could be partially closed off for privacy by curtains. Each bungalow or set of rooms had its own dispenser of filtered 'drinking' water and glasses.

Shower and toilet facilities were located a short distance from each bungalow in separate buildings. Showers usually consisted of 3-6 separate, unscreened rooms in a raised, thatched-roof building. Water was pumped from the *quebrada* to a holding tank and was usually available 24 hours a day (except when the tank was being refilled). Toilet facilities were latrines consisting of a box-like platform with a toilet seat attached around an opening, the entire structure of which was located over a pit. Three to four such rooms made up each latrine facility, which were also separate, covered, unscreened buildings. All bungalows, showers, latrines, and other facilities were connected by means of a covered walkway.

Two open-air hammock 'houses' had been constructed to provide a comfortable area for visitors to relax. One was a wide covered passageway furnished with hammocks, tables, and chairs that was located between a bungalow and a screened, gazebo-like room. The other was a square, patio-like arrangement with a high, thatched roof over tables, chairs, and hammocks.

There was no electricity at the Explorama Lodge. Illumination was provided by kerosene lanterns which were filled, cleaned, and lit punctually every day. Each room had one lamp inside, while lamps were placed in holders on the outside walls of the bungalows between rooms. Ground level kerosene 'torches' were also used to light up the outer porches and walkways.

Trail System

There were several distinct, well-used trails at the Explorama Lodge that led out from the facilities into the various habitats of the surrounding forest for a total of 10-12 kilometers. Other, less often used paths brought the total to approximately 48km. Trails were not marked and trail maps were not made available, as visitors were prohibited from walking the trails without a guide, unless special arrangements had been made.

Explorama Tours normally employed 8-10 full-time, bilingual Peruvian guides, whose services are provided to guests regardless of the size of their group. Except under special circumstances, these guides always accompanied tourists on excursions. Once assigned to a group, a guide would remain with that group throughout their entire stay at the different lodges of Explorama Tours.

Costs

Explorama Tours offers a variety of 'packages' to visitors, which include combination stays among the three facilities they operate. The exact charge is dependent upon the length of stay, the number of people in one's group, and the lodge(s) selected to visit. For example, based on information obtained in January 1989, the cost for one or two persons to spend three nights at the Explorama Lodge and three nights at the Explorama Inn was $420 per person. The cost of spending an additional night at the Explorama Lodge is $60. All Explorama programs included airport reception in Iquitos, transfers, land and river transportation in their own vehicles, accomodations, meals, and guided excursions. Visitors were not required to pay any additional charge for the services of guides, but could tip according to personal preferences.

Comments

(See Comments On Explorama Facilities on page 36.)

Explorama Inn

Owner

Peter Jenson

Address

Explorama Tours
Box 446
Iquitos, Peru
Telephone: 23-34-81 or 23-54-71
Telex: 91014
Fax: 51-94-23-49-68

U.S. Representatives:

> Selective Hotel Reservations
> Telephone: (800)-223-6764 (USA)
> (617)-581-0844 (Inside Massachusetts State)

Description

The Explorama Inn is one of three tourist lodges owned and maintained by Explorama Tours. The Explorama Inn was constructed in 1985.

Location

The Explorama Inn is located on the banks of the Amazon River near the village of Indiana, northeast of Iquitos in the department of Loreto. It is approximately 40km by river from Iquitos to the Explorama Inn. The Inn is situated between 3°-4° South latitude at an elevation of approximately 140 meters.

Logistics

The Explorama Inn is easily reached from the Amazon port city of Iquitos, which is also the capital of Loreto Department. Both the Peruvian airlines Faucett and AeroPeru have regularly scheduled flights between Lima and Iquitos. Faucett has also maintained a direct flight between Miami and Iquitos, and Iquitos to Miami.

Employees of Explorama Tours met visitors at the Iquitos airport and then used their own bus to drive them 15-20 minutes to the boat dock. The administrative offices of Explorama Tours were located at street level above the dock. Typically an outboard motor-powered, thatched-roof boat about 10 meters in length and capable of holding 25-30 people is used for the river trip. It took approximately 1 $^1/_2$ hours to travel down the Amazon River from Iquitos to the Explorama Inn. The same distance on the return trip against the current took about 2 hours.

Passengers disembarked along the banks of the Amazon and walked up a slight hill to the inn. Suitcases and heavier items were carried by Explorama Inn employees.

Forest Type

Forest at the Explorama Inn is classified as lowland Tropical Moist Forest. It is mostly on relatively rich clay soil and constitutes a small and somewhat disturbed forest island.

Seasonality

The area around the Explorama Inn receives approximately 3,050mm of rain per year. Every month of the year (except May) has been the wettest, and most months have been the driest of at least one calendar year during the past three decades for which records are available. The Amazon River water levels are highest in the Iquitos region usually during the months of December through June, and lowest from July through November.

Facilities

The Explorama Inn is located on 38 hectares of privately-owned and protected land. The main building combined a lounge, dining room, and kitchen facilities, which were constructed on a level area above the river bank and flood zone. Outside the inn was a large, tiled patio, approximately half of which was covered by a high, thatched roof equipped with overhead fans. Landscaped with tropical plants, lounge chairs were furnished for visitors who wished to take in the sun, air, or scenery.

The first portion of the Inn one entered was the lounge. This was entirely screened and provided with overhead fans and electric lights. A bar ran the length of one side where soda, beer, and liquors were sold. There was also a dispenser of drinking water available to all guests, located on the bar. Chairs and crude tables of sawn tree sections could seat approximately 40-50 people. Native Indian handicrafts were on display and for sale.

Continuing from the lounge to the dining room, one passed an open porch or breezeway furnished with chairs and table. A long hammock house reached from this breezeway by a set of stairs, was available for relaxation. It was unscreened and covered with a thatched roof.

A guest cabin at the Explorama Inn. (Near Iquitos, Peru)

The dining room itself was screened and equipped with overhead lights and fans. It could accomodate about 70 people. Food was laid out on long tables at one corner of the room with buffet-style service. Food at the Explorama Inn was excellent. Fresh fruits and fish were obtained locally from the Iquitos area, while fresh vegetables and other items were shipped in from Lima. Thermos jugs of hot water were available, along with tea, instant coffee, and instant hot chocolate, which guests had access to 24 hours a day.

The Explorama Inn could accommodate approximately 52 people. Guests were housed in private thatched-roof cabins constructed on cement slabs. A total of 26 cabins were situated among tropical foliage to provide maximum privacy. Cement walkways offered access among the cabins and between the cabins and the Inn. Cabins were screened and typically had two single beds, electric lights, an electric fan, shelves, and hangers. There was also a shower, flush toilet, and sink in the cabin, separated from the beds by a wall-like partition. Shower and sink water came from a well, but was not potable. A pitcher of purified drinking water was placed in every cabin and refilled daily.

The Explorama Inn had a generator that normally produced electricity 24 hours a day. The electricity was 220V with combination outlets in the rooms that took a plug with two round shaped prongs or two flat, parallel prongs.

Trail System

There were several trails at the Explorama Inn that offered access to the surrounding forest. A total of 20km of cut trails that were unmarked, and for which trail maps were unavailable extended throughout the privately owned land. Visitors were prohibited from walking the trails unless accompanied by a guide.

Explorama Tours normally employed 8-10 full-time, multilingual Peruvian guides, whose services were provided to guests regardless of the size of their group. Except under special circumstances, these guides always accompanied tourists on excursions. Once assigned to a group, a guide would remain with that group throughout their entire stay at the different lodges of Explorama Tours. Some of the guides have worked with reknowned biologists and ornithologists and are capable of imparting a wealth of information to visitors.

Costs

Explorama Tours offers a variety of 'packages', which include combination stays among the three facilities they operate. The exact charge is dependent upon the length of stay, the number of people in one's group, and the lodge(s) selected to visit. For example, based on information obtained in January 1989, the cost for one to two persons to spend three nights at the Explorama Lodge and three nights at the Explorama Inn was $420 per person. Each additional night at the Explorama Inn cost $50. All Explorama programs included airport reception in Iquitos, tranfers, land and river transportation in their own vehicles, accomodations, meals, and guided excursions. Visitors were not required to pay any additional charge for the service of guides, but could tip according to their personal preferences.

The owner of Explorama Tours, Peter Jenson, has a history of cooperating with scientists that wish to conduct long-term studies at his facilities. Biologists with legitimate research projects are advised to contact Mr. Jenson regarding specific needs and prices.

PERU

Comments

(See Comments On Explorama Facilities on page 36.)

Explornapo Camp

Contact

Peter Jenson

Address

Explorama Tours
Box 446
Iquitos, Peru
Telephone: 23-34-81 or 23-54-71
Telex: 91014
Fax: 51-94-23-49-68

U.S. Representatives:

Selective Hotel Reservations
Telephone: (800)-223-6764 (USA)
(617)-581-0844 (Inside Massachusetts State)

Description

The Explornapo Camp is one of three tourist facilities owned and maintained by Explorama Tours. It was constructed in 1982 to provide accomodations in the Amazon Basin at a remote location from Iquitos.

Location

The Explornapo Camp is located in the department of Loreto, approximately 90km northeast of the city of Iquitos. The actual camp is situated near the banks of the Sucusari River, a tributary of the larger Napo River that flows down from Ecuador.

The Camp was constructed at an elevation of approximately 140 meters and is located between 3°-4° South latitude.

Logistics

To visit the Explornapo Camp one must first reach the city of Iquitos. Both the Peruvian airlines Faucett and AeroPeru have regularly scheduled flights between Iquitos and Lima. Faucett also offers a direct flight from Miami to Iquitos, and Iquitos to Miami as well.

Arriving guests were met by representatives of Explorama Tours at the Iquitos airport. Transportation was then provided by private bus for the 15-20 minute ride from the airport to the boat dock on the Amazon River. Located directly below the administrative offices of Explorama Tours, one of their fleet of boats took guests to the Explornapo Camp. The most common type of boat used was an outboard motor-powered, thatched-roof vessel of about 10 meters length and capable of holding 25-30 people.

The journey by boat covered approximately 160km along rivers. From Iquitos, it traveled down the Amazon River to the confluence with the Napo River. There the boat turned and headed up the Napo River in a generally northwesterly direction. The boat remained on the Napo until entering the Sucusari, which was only traveled for 5-10 minutes before arriving at the floating dock. The campsite was constructed on high ground just above the dock and beyond the high water level.

The river trip from Iquitos to the Explornapo Camp took about 8 hours. The return journey was normally $6\,^1/_2$ hours. However, visits to the Explornapo Camp were usually coordinated with at least the first and last nights spent at the Explorama Lodge. The trip by boat from the Explorama Lodge and the Explorama Inn to the Explornapo Camp was $5\,^1/_2$ and $6\,^1/_2$ hours, respectively.

Forest Type

The forest in the area of the Explornapo Camp is classified as lowland Tropical Moist Forest. Both seasonally inundated and non-inundated forest types are present near the Camp. This area is on the north bank of the Napo River and apparently contains some plant and animal taxa that do not cross that river in the immediate area of Iquitos.

Seasonality

The Explornapo Camp site is generally aseasonal. Rainfall normally ranges between 3,000-4,000mm, with records indicating that every month normally receives a minimum of 170mm. High water months are generally from December through June and low water months from July to November (with respect to river water levels), but may vary greatly.

Facilities

The Explornapo Camp was constructed on the 1,792ha Explornapo Reservation, a private reserve that is owned and maintained by Explorama Tours. The camp was a small complex of elevated, thatched-roof, open air buildings connected to one another by wooden walkways. Also accessible from the Camp were the 125ha Shimigay Reserve and the 4,770ha Orejon Reserve.

The first area of the camp facilities one entered was the dining room. Here, two long tables with crude benches were used to serve visitors. Dining was family style with the food simple, but good. Drinking water was transported by boat from Iquitos and was accessible in a dispenser in the dining area. Tea, instant coffee, and instant hot chocolate were also always available with thermos jugs of hot water. Supplies were usually transported on the same boat as passengers and normally included soda and beer.

From the dining area extended a long open hallway with hammocks that led to a sleeping area. One passed the showers on the way, which used water supplied by the Sucusari. There were two shower stalls, which were covered but unscreened. The sleeping area was a raised, covered, unwalled platform like the majority of the camp structures. Beds consisted of small matress pads laid directly on the structure floor, and covered with

Sleeping quarters at the Explornapo Camp. (Near Iquitos, Peru)

mosquito nets. Visitors usually slept on one side of the sleeping room, while guides and employees slept on the other. About 15 meters beyond the sleeping area was the latrine.

The latrine consisted of four toilet facilities in an unscreened,

The open-air lodge at the Explornapo Camp. (Near Iquitos, Peru)

covered building. Toilet seats fastened to raised boxes located over excavated pits were the toilets.

The Explornapo Camp accommodates groups of up to 40 people. It was limited to this number to prevent detrimental effects to the surrounding forest. There was no electricity. Kerosene lanterns and 'torches' provided all artificial illumination. The facilities were rustic, and are most often used by serious bird watchers and groups wishing to visit more isolated and less disturbed areas of rainforest.

Explorama Tours now offers visitors the opportunity to visit salt 'licks' or *colpas* located near the Explornapo Camp. This provides the possibility of viewing at night certain difficult to see animals. Visits to the *colpa* are made by special arrangement and usually include camping out in *tambos*, small, raised platforms covered with palm leaves and provided with sleeping mats and mosquito nets.

Trail System

There was a series of approximately 100km of unmarked and unmapped trails that interlace through the forest at the Explornapo Camp. Unless prior arrangements have been made, guests were asked to use these trails only while accompanied by a guide.

Explorama Tours normally employed 8-10 full-time, multilingual Peruvian guides, whose services were provided to guests regardless of the size of their group. These guides would always accompany tourists on excursions. Once assigned to a group, a guide would remain with that group throughout their entire stay at the different lodges of Explorama Tours.

Costs

Explorama Tours offers a variety of 'packages' to visitors, which include combination stays among the three facilities they operate. The exact charge is dependent upon the length of stay, the number of people in one's group, and the lodge(s) selected to visit. For example, based on information obtained in January 1989, the cost for one person to spend two nights at the

Explornapo Camp, two nights at the Explorama Lodge, and two nights at the Explorama Inn was $1100. The same program with a group of two people cost $650 per person. Each additional night spent at the Explornapo Camp costs $80. All Explorama programs included airport reception in Iquitos, transfers, land and river transportation in their own vehicles, accomodations, meals, and guided excursions. Visitors were not required to pay any additional charge for the services of a guide, but may tip according to their personal preferences.

The owner of Explorama Tours, Peter Jenson, has a history of cooperating with scientists that wish to conduct long-term studies at his facilities. Biologists wih legitimate research projects are advised to contact Mr. Jenson regarding specific needs and prices.

Comments On Explorama Facilities

Explorama Tours is one of the best organized and most well run companies that I have had contact with. I base this on over three months of conducting research at all three of the Explorama facilities. Owner Peter Jenson lives in Iquitos where he can supervise operations at the three Explorama lodges from his administrative office overlooking the Amazon River. His 25 years of experience gained from running the Explorama Lodge have been applied to the newer Explorama Inn and Explornapo Camp. The result is a smooth running operation that causes minimal inconvenience to visitors. Greatly aiding this operation is his crew of 80 employees, including multilingual guides, his fleet of 8-10 river boats, and an excellent business relationship with Faucett Airlines, as well as the hotel and airport staff in the city of Iquitos.

The three lodges themselves can accomodate guests in varying degrees of comfort and offer something for everyone. The well established Explorama Lodge has no electricity, but provides comfortable rooms in its guest bungalows. Access to the forest is easy along the clearly defined trails. Excursions from the Explorama Lodge can include visiting a nearby schoolhouse, as well as a local tribe of indigenous Yagua Indians. Although acculturated, the Yaguas will dress in traditional garb and exhibit native handicrafts for visitors. The Explorama Inn is extremely comfortable and logistically simple to reach from Iquitos. This facility is highly suggested for those with only 1-2 days to spend at a location, or for those preferring more 'civilized' accomodations. The separate cabins provide maximum privacy for those uncomfortable with the shared accomodations of a bungalow. Excursions included guided walks in the forest, as well as visits to nearby lakes to see water fowl and giant water lilies. One can also hike to a large nearby village (Indiana) and take a boat ride to visit the Yagua Indians.

The rustic Explornapo Camp is primarily for serious outdoor enthusiasts, birdwatchers, or field biology groups. It is the least accessible facility with the fewest comforts, but also with the distinct characteristic of showing the least amount of disturbance due to human influence. One's best chance of seeing a variety of monkeys and large mammals is at the Explornapo Camp, whereas large mammals and birds have been essentially eliminated from forest nearer Iquitos. Excursions include traveling by boat along a black water tributary and searching for primitive hoatzin birds at an oxbow lake. Although the living conditions are more rugged, it should be easily tolerable to visitors used to camping.

The staff of Explorama Tours was always friendly and willing to help. The management and guides were eager to accomodate visitors with special interests, although it helps if they have advance knowledge of special ecursions desired by guests. Peter Jenson has aided biologists in the past with logistical support and discounted rates. Therefore, legitimate researchers planning to use any of the Explorama lodges as a base for long-term studies should contact Mr. Jenson at the address provided earlier.

The rainforest near Iquitos has been subject to heavy pressure from the 450,000 inhabitants of the city and surrounding area. Many large trees (especially Ceiba) have been logged out while larger animals have been consistently hunted. However, the number of resident bird species remains high, as does the diversity of trees and plants in the area.

Facilities In The Iquitos Area

The Iquitos area of northeastern Peru has more 'jungle camps and lodges' than any other area in Latin America that I have visited. Facilities are usually located along the Amazon River or one of its tributaries, within a day's journey of Iquitos.

They offer visitors access to the rainforest with various degrees of comfort. The offices and agents for the majority of these enterprises are located near or around the *plaza de armas* in Iquitos (within 1-2 blocks of the Hotel de Turistas).

The following is a partial list of facilities and their agents I compiled from brochures and other printed matter obtained in Iquitos. A particularly useful brochure is produced by FOPTUR (Fondo De Promoción Turistica) with the title of *Albergues Y Campamentos Turisticos De Selva - Departamento De Loreto - Perú* (Jungle Lodges And Tourist Camps - Department of Loreto - Peru).

Amazon Botanical Gardens
P.O. Box 105
Iquitos, Peru
Telephone: (816)-931-2840 (For reservations in the U.S.)

The Amazon Botanical Gardens is a newly constructed lodge built on the banks of the Amazon River, approximately 56km downriver from Iquitos.

Amazon Camp Tourist Service
Prospero 151
Iquitos, Peru
Telephone: 23-39-31 or 23-40-07

The Amazon Camp Tourist Service conducts excursions to four tourist facilities near Iquitos. The lodge Amazon Camp is situated on the Momon River, a tributary of the River Nanay. It is about 25km northwest of Iquitos. The Tambo Yarapa is approximately 80km south of Iquitos and about 15km northeast of Nauta. It is located on the Yarapa River. The Tambo Tahuayo is located on the Huaisi River, a small branch of the Tahuayo River. Tambo Tahuayo is approximately 50km south of Iquitos. The fourth facility is the Madre Selva located on the Apayacu River near Lake Atuncocha, approximately 120km northeast of Iquitos. The Amazon Camp Tourist Service also offers cruises of varying lengths down the Amazon River. Some of these cruises go as far as the Brazilian and Colombian borders.

Amazonia Expeditions, Inc.
Paul Beaver and Millie Sangama
1824 N. W. 102nd Way
Gainesville, FL 32606 USA
Telephone: (904)-332-4051

Amazonia Expeditions, Inc. wholesales and retails expeditions out of the Iquitos area. They visit the Yarapa River, the Tahuayo River, and the Pacaya-Samiria National Park.

Amazonia Expeditions, Ltda.
Carlos Grandez
Putumayo 139
Iquitos, Peru
Telephone: 23-63-74
Telex: 25061

Amazonia Expeditions, Ltda. operates excursions based out of a facility located on the Yarapa River.

Amazon Selva Tours S.A.
133 Putumayo
Iquitos, Peru   OR
Telephone: 23-69-18
Telex: 91002 PE PB IQUIT

Wilderness Expeditions
310 Washington Ave., S.W.
Roanoke, Virginia 24016 USA
Telephone: (703)-342-5630
(800)-323-3241
Telex: 901238

Amazon Selva Tours/Wilderness Expeditions conducts programs from three tourist facilities that offer access to the rainforest. The Anaconda Lodge is located about 40km west/northwest of Iquitos on the Momon River. The Tambo Safari Inn is located approximately 64km southeast of Iquitos on the Tamshiyacu River. The Tambo Safari Camp is about 100km south of Iquitos on the Yarapa River.

Amazon Lodge And Safaris S.A.
Putumayo 165
Iquitos, Peru
Telephone: 23-30-32
Telex: 91026

Amazon Lodge And Safaris runs tours from two tourist facilities. The Amazon Lodge is situated approximately 20km northeast of Iquitos on the Yanayacu River. The Campo Curaca is on the Maniti River, about 25km east of Iquitos.

Explorama Tours
(Exploraciones Amazonicas S.A.)
Avenida La Marina 340
Iquitos, Peru
Telephone: 23-54-71   Telex: 91014

Mailing Address:
Box 446
Iquitos, Peru

Explorama Tours operates three tourist facilities: the Explorama Lodge, the Explorama Inn, and the Explornapo Camp(all of which have been described previously on pages 21-37.).

Paseos Amazonicos Ambassador
Pevas 246
Iquitos, Peru
Telephone: 23-31-10
Telex: 91086

Paseos Amazonicos Ambassador operates programs from the Sinchicuy Lodge located on the Sinchicuy River, approximately 25km north of Iquitos.

The following companies or enterprises only offer programs at their own lodges and facilities.

Hotel Amazonas Plaza
Avenida Aberlado Quinones, KM 3.5
Iquitos, Peru
Telephone: 23-10-91
Telex: 91055

The Hotel Amazonas Plaza operates the Amazon Village and Club located on the Momon River, approximately 20km west/northwest of Iquitos.

Jungle Amazon Inn S.A.
Putumayo 132
Iquitos, Peru
Telephone: 23-22-49
Telex: 91030

The company Jungle Amazon Inn maintains the lodge called Jungle Amazon Inn approximately 30km northeast of Iquitos on the Amazon River at Timicuro.

Amazon River Lodge
Putumayo 184
Iquitos, Peru
Telephone: 23-39-76

The company Amazon River Lodge operates the Tamshiyacu Lodge about 25km southeast of Iquitos on the Amazon River.

Lima

Most visitors to Peru will start in Lima. The Jorge Chavez Airport is fairly large and modern, with numerous shops, a cafeteria upstairs, and a coffee shop on the ground level. There is also a bank where you can exchange currency and an information counter (with bilingual personnel) where you can obtain information regarding hotels and hostals. One of the nicest features of the airport is a 24 hour luggage storage room/service (called the *Guardiania*). The price per item (regardless of size) was $.50 per hour or $2.50 per day. I have used this service often, and have never had a problem with theft or rifling of bags, even when items of obvious value (like cameras) were stored.

While in Lima, I usually stay in the Miraflores area. This is a district of residences, shops, and restaurants, and is generally thought of as being safe. A taxi ride from the airport to Miraflores should take about 20-30 minutes and cost no more than $5. Even five dollars is steep, but cab drivers will certainly try to exercise the 'gringo factor' to their advantage. Therefore, it is always wise to agree on a price before getting into a cab.

Miraflores has a variety of luxury hotels, as well as inexpensive *hostales* and *pensiones*. I usually stay at the Hostal Señorial, which is located on Calle Jose Gonzales, about a block from Avenida Larco and 2-3 blocks from the Grand Hotel Miraflores. This hostal has 28 rooms on three floors and is capable of accomodating 40-50 people. Rooms look out onto a gardened courtyard at the rear of the building. There is a locked gate at the street and then a locked door at the front entrance. Both are operated by a buzzer controlled from the desk, which has a view of the street and is maintained 24 hours a day. Usually someone bilingual mans the front desk.

The Hostal Señorial is frequently used by tour groups, so it is advisable to make a reservation before arrival. A 'single' cost $22/night and included a modern room with two double beds, closet, and bureau. The private bathroom had sink, toilet, and shower with 24 hour hot water. A light breakfast was also included with the price of the room.

Hostal Señorial
Jose Gonzales 567
Miraflores, Lima, Peru
Telephone: 45-97-24 or 45-73-06

Another hostal I've stayed at in Lima is the Hostal Roma on Calle Ica in downtown Lima. This is not a safe area (due to thieves and robbers), and I only used it as an alternative to spending five hours alone in the airport late at night. The cost was approximately $2.50/night for a small room which had shared toilet and bathing facilities outside.

Note: All prices quoted were in effect during February 1989. Peru currently suffers from one of the world's highest inflation rates.

Cuzco

Cuzco is an interesting city situated high in the Andes at an elevation of about 3,300 meters. It attracts thousands of tourists each year on their way to visit Machu Picchu. I highly recommend at least one visit to Cuzco in order to take in the Incan ruins of Machu Picchu and the Sacred Valley. Since flights between Lima and Puerto Maldonado generally make an obligatory stop in Cuzco, it is a simple matter for those visiting areas in southeastern Peru to set aside a few extra days in Cuzco. During my last trip to Cuzco, I stayed at the following inexpensive hostal:

Hostal Residencial
Leonard's Lodgings
Avenida Pardo 820
Apartado 559
Cusco, Peru
Telephone: 23-28-31

This hostal is a private residence, fenced-in from the street, with a number of rooms that are rented out. The room I had held a double bed, desk, chest of drawers, and bathroom with sink, toilet, and hot water shower (although water was only available during certain hours). The cost of the room for just myself as a single was $10/night, with a light breakfast included.

The hostal Leonard's Lodgings is located on Avenida Pardo near the intersection of Avenida Garcilaso. Almost directly in front of the hostal is a traffic circle with a statue in the center. This hostal is one block from Avenida El Sol (which runs parallel to Avenida Pardo), and approximately 6-7 blocks from the Plaza de Armas. The office of Manu Nature Tours (627-B Avenida El Sol) is approximately three blocks from the hostal Leonard's Lodgings. To reach it, one goes straight from the hostal down Avenida Garcilaso and then makes a left at Avenida El Sol (the first intersection). One walks up the hill and the three story building is on the right hand side. The offices for the Conservation Association for the Southern Rainforest of Peru are in the same building.

Another hostal that has been recommended to me by biologists is the Hostal Machu Picchu. It is near the Plaza de Armas across from the old AeroPeru office, about two blocks up from the Avenida El Sol. It is rather spartan, but very inexpensive ($3/night) with an attractive colonial style central court yard. The owner/manager has a tradition of helping visiting biologists by providing equipment storage and other services.

One should exercise caution while sightseeing and walking along the streets of Cuzco. I had my pocket 'picked' in the area between the Plaza de Armas and the train station. This was deftly done in broad daylight on a crowded street. I believe it was accomplished by 2-3 women working together. One should not carry money, credit cards, or other items of value that are not essential to have on your person.

Note: All prices quoted were in effect during January 1989.

Comments On Peru

Peru is one of the most beautiful countries in Latin America and has something for everyone. Tourism is much more highly developed, due primarily to Machu Picchu. 'Jungle tourism' is at its peak in the area around Iquitos, but is also present at numerous other sites in eastern Peru. In the southeastern department of Madre de Dios one finds one of the most biologically rich and least disturbed areas of the Neotropics.

Roads into the forested eastern regions of the country are basically lacking; the most widely used modes of transportation being boats or airplanes. The two Peruvian airlines (AeroPeru and Faucett) provide numerously scheduled flights throughout the country, but are notorious for their delays and cancellations.

Food and lodging is quite inexpensive in most Peruvian cities. Beautiful *artesania* in the form of tapestries and hand-painted ceramics are available to visitors. However, thievery and muggings occur, so that one should be careful as to where they walk. A group of Maoist guerillas called the *Sendero Luminoso* (Shining Path) causes problems periodically, primarily in the area near Ayacucho. The biggest national problem however, is the cultivation of cocaine in the Huallaga River Valley and especially around Tingo Maria. Conflicts between drug traffickers and troops attempting to eradicate the coca fields and processing labs have made that area of Peru particularly dangerous.

Books

TALES OF THE PERUVIAN AMAZON
Beaver, Milly Sangama de and Paul Beaver
1989. AE Publications

LOST CITY OF THE INCAS - THE STORY OF MACHU PICCHU AND ITS BUILDERS
Bingham, Hiram
1948. Duell, Sloan and Pearce

BACKPACKING AND TREKKING IN PERU AND BOLIVIA
Bradt, Hilary
1987. Bradt Enterprises

LOS PARQUES NACIONALES DEL PERÚ
Dourojeanni, Marc and Carlos F. Ponce
1978. Instituto De La Caza Fotografica y Ciencias De La Naturaleza

ADVENTURING IN THE ANDES: THE SIERRA CLUB TRAVEL GUIDE TO ECUADOR, PERU, BOLIVIA, THE AMAZON BASIN, AND THE GALAPAGOS ISLANDS
Frazier, Charles with Donald Secreast
1985. Sierra Club Books

EXPLORING CUZCO
Frost, Peter
1989. Nuevas Imagenes

FOUR NEOTROPICAL RAINFORESTS
Gentry, Alwyn (Editor)
1990. Yale University Press

ROUGH GUIDE TO PERU
Jenkins, Dilwyn and Clare Jenkins
1985. Routledge Chapman and Hall

WIZARD OF THE UPPER AMAZON - THE STORY OF MANUEL CORDOVA RIOS
Lamb, F. Bruce
1974. North Atlantic Books.

RIO TIGRE AND BEYOND - THE AMAZON JUNGLE MEDICINE OF MANUEL CORDOVA
Lamb, F. Bruce
1985. North Atlantic Books

THE PEOPLES AND CULTURES OF ANCIENT PERU
Lumbreras, Luis (Translated by Betty J. Meggers)
1974. Smithsonian Institution Press

A TRAVELER'S GUIDE TO EL DORADO AND THE INCAN EMPIRE
Meisch, Lynn
1984. Penguin Books, Inc.

PERU: A TRAVEL SURVIVAL KIT
Rachowiecki, Rob
1987. Lonely Planet Publications

MICHAEL'S GUIDE: BOLIVIA AND PERU
Shichor, Michael
1987. Inbal Travel Information Ltd. (Israel)

FIVE NEW WORLD PRIMATES - A STUDY IN COMPARATIVE ECOLOGY
Terborgh, John
1983. Princeton University Press

CUT STONES AND CROSSROADS: A JOURNEY IN THE TWO WORLDS OF PERU
Wright, Ronald
1984. Viking Press

Maps

There is a limited selection of both topographical and street maps available in bookstores, such as the ones found near the Plaza de San Martin in downtown Lima. High quality topographical and departmental maps can be obtained at the Instituto Geográfico Nacional on Avenida Aramburu in the San Isidro district of Lima. Domestic sources of Latin American maps are listed on pages 290 and 291.

Tourist Information Sources

Dirección General De Turismo
Ministerio De Industria, Comercio, Turismo E Integración
Calle 1 Oeste, Corpac
San Isidro, Lima 27, Perú
Telephone: 40-71-20
Telex: 20194

Empresa Nacional De Turismo (ENTURPERU)
Avenida Javier Prado Oeste 1358
San Isidro, Lima, Perú
Telephone: 40-46-30
Telex: 20393

Fondo De Promoción Turistica (FOPTUR)
Avenida Republica De Panamá 3055, 16 piso
San Isidro, Apartado 100-56
Lima 27, Perú
Telephone: 41-97-78
Telex: 21363

South American Explorers Club
Casilla 3714
Lima 100, Perú
Telephone: 31-44-80

or

South American Explorers Club
P.O. Box 18327
Denver, Colorado 80218 USA
Telephone: (303)-320-0388

Tambopata Nature Tours
Avenida Sol 627-B
Oficina 202
Cusco, Perú
Telephone: 23-47-93

Conservation Organizations

Asociación Peruana Para La Conservación De La Naturaleza (APECO)
Parque José Acosta 187, 2 piso
Magdalena - Lima 17, Perú
Telephone: 61-63-16

Asociación De Conservación Por La Selva Sur Del Perú
Portal De Panes 137
Plaza De Armas
Cusco, Perú
Telephone: 23-47-93

or

Conservation Association For The Southern Rainforest Of Peru
228 Thompson Mill Road, RD 2
Newton, Pennsylvania 18940 USA
Telephone: (215)-598-7047

Asociación AMETRA 2001 (Aplicación De Medicina Tradicional)
Casilla 42
Puerto Maldonado
Madre De Dios, Perú

PERU

Scientific Organizations/Institutions

Sociedad Entomológica Del Perú
Apartado 4796
Lima 100, Perú

Instituto Nacional De Investigación Agraria
(National Agricultural Research Institute)
Sinchi Roca 2728-Lince
Apartado 2791
Lima 14, Perú

Museo De Historia Natural 'Javier Prado'
Universidad Nacional Mayor De San Marcos
Avenida Arenales 1256
Apartado 140434
Lima 14, Perú

Universidad Nacional Agraria
Apartado 456
La Molina
Lima, Perú
Telephone: 35-20-35

Universidad Nacional De La Amazonia Peruana
Apartado 496
Iquitos, Loreto, Perú
Telephone: 23-53-51

Universidad Nacional Agraria De La Selva
Apartado 156
Tingo Maria, Huanuco, Perú
Telephone: 2341

CHAPTER 1B ECUADOR

General Information

Area: 271,000 sq km (104,500 sq miles)

Population: 10,490,000

Capital: Quito (population: 1,200,000)

Language: Spanish

Currency: Sucres (1 sucre = 100 centavos)
Coins: 20 centavos, 50 centavos, and 1 sucre.
Notes: 5, 10, 20, 50, 100, 500, 1,000, and 5,000 sucres.
Exchange Rate: U.S. $1 = 500 sucres (October 1988)
U.S. $1 = 700 sucres (March 1990)

National Airlines: Ecuatoriana, TAME

Telephone Codes: Ecuador: 593
Quito: 2
Guayaquil: 4

Ecuador

1 - Tinalandia   2 - Rio Palenque Science Center   3 - La Selva Lodge
4 - Jatun Sacha Biological Station   5 - Aliñahui Cabins

Introduction

Ecuador is the fourth smallest republic in South America, approximately the size of the state of Colorado. However, it supports a wide variety of ecological habitats due to its varied terrain. Located on the western coast of South America and encompassing the equator, Ecuador is bordered by Colombia to the north, by Peru to the south and east, and by the Pacific Ocean to the west. The Galapagos Islands are also a part of Ecuador, and are located approximately 1,000 km (600 miles) to the west of the mainland.

Ecuador is divided into three main geographical regions. The Andes Mountains cut through the center of the country from north to south in two primary ranges called the Eastern and Western Cordilleras. The area between them is known as the Central Valley. To the west of the Central Valley is the coastal region and main agricultural area of Ecuador. While to the east of the Central Valley is the vast region of lowland rainforest called the Oriente. Although the Oriente comprises more than a third of Ecuador's total land area, it has traditionally remained lightly populated. The discovery and exploitation of petroleum reserves in this region however, has resulted in an increase in its population.

My evaluation of rainforest sites in Ecuador took place during October 1988. During this time I visited two sites on the western slopes (Tinalandia) and foothills (Rio Palenque Science Center) of the Andes, and two sites in the eastern lowlands or Oriente region(La Selva Lodge and Jatun Sacha Biological Station). Information is also presented about the Aliñahui Cabins, the Maquipucuna Tropical Reserve, and the Rio Guajalito Reserve.

Tinalandia

Contact

    Tina and Alfredo Garzon

Address   Cable

    Tinalandia   Hotel Tinalandia
    Santo Domingo de Los Colorados   Santo Domingo de Los
    Ecuador   Colorados
    (No telephone)   Casilla 8
                                          Ecuador

Description

    Tinalandia was a resort hotel, some of its facilities having originally been constructed in the 1950's. The 9-hole golf course, as well as the architecture of the swimming pool and dining hall, suggest the style of Old World aristocracy. Groups of cottage-like cabins of recent construction are available to house visitors that come mainly to watch birds and collect insects.

Location

    Tinalandia is located in Pichincha Province of northwest Ecuador, approximately 12km east of the city of Santo Domingo de los Colorados. It is situated between 0°-1° South latitude. Situated on the western slopes of the Andes, Tinalandia formerly encompassed 2,000 acres of land planted primarily in sugar cane and coffee. Today, Tinalandia consists of approximately 600 acres, which has the status of a wildlife preserve. Of this land, a 60 acre tract is primary forest that has never been cut. Located at an elevation of about 2,000 feet, about 150 acres of land is located on the north bank of the Rio Toachi, while the remaining acreage (including the primary forest and living and dining accomodations) are located south of the river.

Logistics

    To reach Tinalandia from Quito, one can inexpensively take the bus that goes to Santo Domingo. You should alert the driver that you wish to be let off at the Hotel Tinalandia. This should cause no problems, as there is a bus stop at the Tinalandia driveway. (Note: The bus stop at the base of Tinalandia is on the opposite side of the road as the driveway and happens to be located at a particularly treacherous stretch of road. Be extremely careful while crossing the highway!) The driveway or entrance road is on an upward slope and extends for approximately 0.2km until the Tinalandia dining hall/restaurant is reached.

When traveling to Tinalandia from Quito by car, one first needs to get onto Avenida Occidental and head towards the southern portion of Quito (Quito Sur). Continuing south, you pass through three separate tunnels and then continue along the main road. The next major road needed is the Panamericana Sur, which unfortunately is very poorly marked. I reached it about 10-15 minutes after passing through the last tunnel, in moderate to slow traffic. It was preceeded by a traffic circle, which I followed $3/4$ of the way around, and shortly thereafter saw the only 'Panamericana Sur' road sign. Staying on this road where the sign was seen, led to a T-intersection. Here a right turn was made onto the Panamericana Sur. At this point heavy truck and bus traffic began and continued all the way to Tinalandia.

One remains heading south on the Panamericana Sur and after about 25km will pass through the town of Tambillo. Approximately 7km further south along the same road will be the town of Aloag. Just before reaching Aloag, there will be a turnoff for 'Rt. 30/Santo Domingo' on the right. Take this right hand turn and Santo Domingo will be located approximately 102km from this point. Once on Rt. 30, one will continue to travel northwest until reaching Tinalandia. The small town of La Palma will be passed approximately 72km after entering Rt. 30. Tinalandia is about 12km southeast of Santo Domingo, and will be reached before the town itself when traveling from Quito.

A stone reminiscent of a gravemarker and carved with the name 'Tinalandia' is located on the left side of the road, immediately adjacent to the Tinalandia driveway. This driveway leads uphill, approximately parallel to Rt. 30. It is easy to enter when traveling northwest on Rt. 30 (from Quito), although one must cross traffic to reach it. (Note: This entrance driveway is located at a dangerous stretch of highway and one should be aware of passing trucks and buses as one slows down to look for the Tinalandia entrance.)

Forest Type

The classification of the forest at Tinalandia is Tropical Wet Forest grading into Premontane Wet Forest. This area is especially rich in epiphytes. Most of the forest is rather disturbed, as is most of the remaining forest in central Ecuador.

Seasonality

The wettest months at Tinalandia are October-November and March-April. There is a fairly pronounced dry season that ranges from early May to June, and again in September.

Facilities

The Hotel Tinalandia could comfortably accomodate approximately 40 guests at one time. The dining hall was the lowest (with respect to elevation) building on the property. All meals were taken at the dining hall, although a boxed lunch could be provided with advance notice. The dining area itself consisted of an open-air, L-shaped porch that overlooks a patio and the Toachi River. The food at Tinalandia was excellent.

Guest cabin at Tinalandia. (Near Santo Domingo de los Colorados, Ecuador)

ECUADOR

Hotel dining room at Tinalandia. (Near Santo Domingo de los Colorados, Ecuador)

Living facilities in the form of cabins or cottages were located at two main sites on the property. Five cabins providing a total of nine rooms were located near and to the left of the dining hall, along an ascending path that parallels the entrance drive. Three other cabins were located approximately 0.7km from the dining hall along a steeply ascending road that terminated at the top of the mountainside that Tinalandia is built on. These 'upper' cabins provided an additional 13 rooms.

During my visit, I stayed in the most recently constructed cabin (rooms 26 and 27), so that the following description may not be exactly typical of all other rooms. However, examination of at least two other rooms in other cabins showed that there were no major differences. Each cabin had 110V electricity that functioned 24 hours a day, with outlets that accepted the standard plug with two parallel, flat prongs used in the US. The rooms had both overhead and reading lights.

The room was large and carpeted, containing three beds (two doubles and one single), an armoir for clothing, and a large bathroom with sink, flush toilet, and a shower with hot water. Drinking water was placed in each room in a pitcher on the coffee table. Ventilation was provided by permanently open screened-windows in the door and bathroom, as well as unscreened wooden shutters placed on either side of the large picture window found in each room.

Adjacent to the upper cabins was a 9-hole golf course, which afforded a pleasant area for birdwatchers wishing to take a leisurely stroll. Next to the golf course and near the cabins, located at the highest point on the property, is a combination club house and residence. The owners live on the second floor, while the large (approximately 10m x 25m) ground floor was filled with tables and chairs available for the use of visitors. One entire length of wall was made up of picture windows that overlooked the valley.

Trail System

There were several trails through the forest at Tinalandia. Unfortunately, the trails were not marked and there were no prepared maps of the trails or other facilities at Tinalandia. Entrances to these three trails were at the following locations. First, if one comes up the road from the dining hall to the upper cabins, the newest cabin (rooms 26 and 27) will be found on the right before the gate leading to the golf course and parking area. Turn right before this gate and pass the cabin. If you continue on this road for approximately 0.4km you will reach a lily pond on the right hand side. To the right of this pond and behind it is a well worn path in the grass. By following this path(or skirting the edge of the forest) you will reach the beginning of a trail in 60-70 meters.

Another trail, approximately 0.7km past the lily pond on the same road, began by a water holding tank. This trail began in the forest on the right side of the road about 12-15 meters before the holding tank (which is a sunken cistern covered with corrugated metal). A third trail was found by following this same road, through a pasture area, to its end. The road (or path at this point) terminates near the side of a hill and loops around in a small circle. The trail entrance was straight ahead from the end of the path, in the side of the forest. This point is approximately 0.9km from the lily pond or 1.3km from cabin 26-27.

Trails also existed along the mountainside on the other side of the Toachi River. This area can be reached by crossing Rt. 30 at the bottom of the Tinalandia entrance drive and following a footpath to a suspension bridge. Once over the bridge, there were paths to the left or right that led up through a small coffee plantation and then shortly to the forested side of the mountain. A petroleum pipeline runs parallel to the river and provides an area relatively free of undergrowth along its right-of-way. Paths up into the mountainside were easily seen from the pipeline, as well as small streams that run down.

Costs

My cost for an unshared room such as the one described under the Facilities section was $48/day, including taxes and service charges. All three meals each day were also included in this price. Prices are subject to change however, and one should confirm a price with the owner before arriving. If you intend to stay for a week or more, a package deal booked through a travel agent might be to your advantage.

Comments

The Hotel Tinalandia appears to have been origianlly constructed in the manner of an aristocrat's mansion. Although some of its splendor has faded through the years, it continues to offer both accomodations and cuisine that are luxurious considering the location. Although the surrounding area has been considerably disturbed through agriculture and cattle production, the grounds of Tinalandia still attract an abundant and varied bird population, and afford some of the best insect collecting in the world.

The hotel was poorly attended during my visit. According to the owner, approximately 2-3 tour groups visit Tinalandia each month. The busiest season of the year is during the months of May and June. Since there is no phone at Tinalandia, one must cable or write for reservations.

A final note on the drive to Tinalandia from Quito. When asked for an estimate of driving time to Tinalandia, the car rental agency told me $2^1/2$-3 hours. This may be true if there are no other vehicles on the road, but not when you have to contend with innumerable buses and large trucks. The surface of the road from Quito to Santo Domingo is actually quite adequate, but the road itself snakes and curves its way through the mountains.

There are often rocks in the road, and for this reason I would suggest renting a vehicle with high clearance such as a Trooper. I would allow at least four hours for the trip to Tinalandia from Quito, and I would suggest traveling during the morning if possible, as fog often makes visibility poor later in the day.

Rio Palenque Science Center

Contact

> Dr. Calaway Dodson

Address

> Rio Palenque Science Center
> Casilla 95
> Santo Domingo de los Colorados
> Ecuador
> Telephone: 56-16-46

> Or

> c/o Missouri Botanical Garden
> P.O. Box 299
> St. Louis, MO 63166-0299 USA
> Telephone: (314)-577-5100

Description

The Rio Palenque Science Center (RPSC) is a privately owned field station and forest preserve that was originally established in 1970 as a project of the University of Miami. The RPSC offers living accommodations and access to a small tract of primary forest for biologists and other visitors. The area surrounding the RPSC is heavily used for agriculture, being planted with oil palms, bananas, cacao, and corn. A portion of the RPSC property is also planted in oil palms, rubber trees, and various fruit crops.

Location

The Rio Palenque Science Center is located in the central coastal region of western Ecuador at an elevation of 150-220 meters. It is situated between 0°-1° South latitude. It is approximately 47km south of Santo Domingo de los Colorados and 56km north of Quevedo. The small town of Patricia Pilar is located about 2km north of the RPSC.

Logistics

The Rio Palenque Science Center is easily and inexpensively accessible by bus from Quito, Santo Domingo, Quevedo, or Guayaquil. For visitors wishing to drive their own vehicle from Quito, I have provided directions starting at the Hotel Tinalandia. A review of the logistics of the previous section will provide information on reaching the Hotel Tinalandia from Quito.

To reach the Rio Palenque Science Center from Tinalandia, one drives down the Tinalandia entrance road and turns left onto the main highway (Rt. 30). You will now be heading northwest towards Santo Domingo. After traveling approximately 10.6km, but before reaching Santo Domingo itself, there will be a sign indicating a righthand turnoff for Quevedo. Take this right turn and continue for approximately another 10km and you will see a sign indicating Quevedo and Rt. 25 South on the left. Make this left. A sign with green letters on a white background stating 'Centro Cientifico Rio Palenque' will be on the left side of Rt. 25 to mark the entrance to the RPSC property. This point is approximately 47km south of Santo Domingo, 2km south of Patricia Pilar, and 63.5km away from Tinalandia. Many of the farms along the road have signs indicating how many kilometers south from Santo Domingo they are.

After turning off Rt. 25 onto the RPSC entrance road, one immediately encounters a chain that blocks the way. A caretaker lives just beyond this point on the property and will unlock the chain to let you pass. From this point it is approximately 1.6km to the field station building. One follows the signs posted, driving through oil palms, rubber trees, and finally through forest to reach the clearing where the field station building is located.

Forest Type

The forest covering the grounds at the Rio Palenque Science Center is classified as lowland Wet Forest, constituting one of the last small patches of this forest type in western Ecuador. Much of it is disturbed and there are large thickets of spiny bamboo, but the large edge effect makes for excellent birdwatching. The forest is on rich alluvial soil and is especially noteworthy for the prevalence of epiphytes.

Seasonality

The dry period generally extends from August through November, but with frequent rains at night. The wet season occurs from December through May, with heavy rains also occurring in July. Total annual rainfall at the RPSC is probably greater than 3,000 millimeters.

Facilities

The living accomodations, kitchen, and laboratory are all housed in the second floor of the field station building. The first floor is used by a resident caretaker and his family. Living quarters consisted of six rooms which will sleep four people each in bunkbeds, for a total capacity of 24 visitors at one time. Two of the rooms had their own toilet and shower facilities, while the remaining rooms shared facilities that were located between them. Bathroom facilities included cold water showers, flush toilets, and sinks. Bed linens and towels were provided. Each room had a cabinet where clothes could be hung and possessions locked. There was 110V electricity available throughout the building from approximately 6:00 - 10:00pm, when the generator was run.

There was a large kitchen area at the RPSC, adequately supplied with pots, pans, utensils, can openers, etc. Boiled water was available in plastic milk jugs for drinking. The stove functioned by gas bottles, which the caretaker showed how to operate. Visitors have the choice of bringing their own food and doing their own cooking, bringing their own food and arranging to have the caretaker's wife do the cooking, or arranging with the

The biological station at the Rio Palenque Science Center. (Near Santo Domingo de los Colorados and Quevedo, Ecuador)

caretaker to purchase the food and handle all the cooking. The station rates vary accordingly.

There was a dining area that consisted of three picnic tables, accommodating about eight people each. Adjacent were a sofa and chairs. The dining area could also be used as a seminar room and contained a small library with much information about the RPSC flora and fauna. A small separate lab area with a work bench and running water was located off the kitchen/dining area.

Trail System

The Rio Palenque Science Center contains 180 hectares of land, of which approximately 100 hectares exist as primary forest. Approximately 70 hectares are planted in African oil palms, and the remaining 10 hectares in macadamia nuts and other fruits. The property is 3.3km deep and forms the west bank of the Rio Palenque. A series of approximately three kilometers of mapped trails extend throughout various portions of the property (maps are available from the caretaker or by writing to Dr. Dodson). Several paths for vehicles also offer access to different areas.

Costs

The rates for field station use depended upon the arrangements made for food and cooking. At the time of my visit I was quoted $15/night, if I purchased and prepared my own food. For an additional $2/day, an arrangement could be made to have all the food preparation handled by the caretaker's wife. However, to have the caretaker and his wife both purchase and prepare the food would have cost $30/day. The caretaker's wife also washed laundry for a nominal charge.

Comments

The Rio Palenque Science Center is easy to reach by bus or car. The road beyond Tinalandia remains paved, but levels out so that driving becomes much less hazardous. Driving time from Tinalandia to the RPSC is approximately one hour. The grounds and facilities at the RPSC offer comfortable accomodations and surroundings in which to conduct scientific investigations or merely walk through the forest and watch birds. Over 360 bird species have been seen, as well as 350 different species of butterflies. Information on other animal groups found at the RPSC is available at their library.

The vast majority of research done at the RPSC has been botanical in nature, resulting in it being one of the botanically best-known patches of forest in the Neotropics. In 1978, *A Flora of the Rio Palenque Science Center* was published by C.H. Dodson and A.H. Gentry; approximately 100 of the 1100 plant species recorded from the station were new to science. Dr. Dodson is an orchid specialist and maintains a beautiful collection of orchids at his house on the station grounds.

The ability of visitors to select the type of cooking arrangement that best suits their needs, facilitates working at the RPSC. For those that do their own cooking, simple foodstuffs can be purchased in the nearby town of Patricia Pilar.

The main negative point concerning the Rio Palenque Science Center is that it is essentially a refuge or island of forest surrounded by land in agricultural use. This is apparent when one approaches the station on Rt. 25, and when looking out from the station building at the oil palm groves across the Rio Palenque.

La Selva[*]

Contact

    Eric Schwartz

Address

    6 de Diciembre 2816
    P.O. Box 635
    Suc. 12 de Octubre
    Telephone: 550-995 or 554-686
    Telex: 2653 JOLEZ ED.

Description

    La Selva is a recently established jungle lodge, constructed primarily to accommodate tourists. However, the owners have a strong commitment to conservation and biological research. To serve this end and encourage ecological studies, a field station building has been constructed.

Location

    La Selva is located approximately 100km down the Napo River (in a generally easterly direction) from the city of Puerto Francisco de Orellana, which is more commonly known as Coca. It is situated on the north bank of the Napo River near the northwestern portion of Yasuni National Park in Napo Province. The lodge and actual facilities are constructed on the north shore of an oxbow lake called Garzacocha (which means 'heron lake' in Quichua). La Selva falls between 0°-1° South latitude.

---

[*] This is not the La Selva field station that is operated by the Organization for Tropical Studies, which is located in Costa Rica.

Logistics

The first leg of the journey to La Selva is usually made by taking a 40 minute flight on TAME Airlines from Quito to Coca. However, because the airstrip at Coca was being renovated at the time of my visit, I followed a slightly different route in order to reach the Napo River near Coca. The first part of the trip was a 30 minute flight on a TAME Airlines F-28 jet to Lago Agrio (which is also labeled on some maps as Nueva Loja). La Selva personnel met the flight at the small Lago Agrio airport, handled all the luggage, and had vehicles standing by for transferring passengers.

The next leg of the journey was by car, which almost immediately crossed the Aguarico River by ferry a short distance south of Lago Agrio. We traveled generally south along a gravel road for approximately 88km, which took about two hours. The end point of the ride was a small petroleum town called San Carlos, located on the north bank of the Napo River.

From San Carlos I boarded a covered, motorized, dugout canoe for the leg of the trip that would take us down the Napo River. The boat trip took approximately two hours. San Carlos is somewhat closer to La Selva than Coca, but this part of the river trip will usually take 2-3 hours, depending on the height of the river. (The return river journey from La Selva to San Carlos took four and a half hours going against the current.) We disembarked on the north bank of the Napo River, approximately 100km east of Coca.

After walking north for several minutes, an elevated boardwalk is reached that leads through the forest. A walk of approximately 20 minutes along this raised trail terminated in a marshy area where canoes were moored. The luggage had been carried by La Selva employees and was stowed in the canoe. It takes approximately 10 minutes to reach the La Selva dock by paddling the canoe from this point across Lake Garzacocha.

In October 1989, I was informed that the Coca airport had been opened. From Coca, it is a five minute ride by road to the Napo River dock. By boat, it then normally takes 2 hours and 10 minutes to reach Lake Garzacocha and approximately $3-3\,1/4$ hours to travel the return trip.

Forest Type

The forest in the area surrounding La Selva is considered Tropical Moist Forest.

Seasonality

The wettest period of the year in the La Selva area is from May through July, with June being the wettest month. It becomes gradually drier following July until the driest period from November through January is reached. December is the driest month of the year. (These observations are based on the approximately two years that the La Selva site has been occupied.)

Facilities

La Selva had 15 cabins with two single beds each, plus one family cabin, providing accomodations for a maximum of about 35 people. Cabins were built on a slight hill overlooking the lake and the edge of the forest. Each cabin was built on stilts, had a thatched roof and small veranda, and was connected to the other buildings by a raised bamboo walkway. The typical cabin held two single beds, each fitted with a light mosquito netting. Cabins were screened, but the split wooden siding of the native style architecture left cracks large enough for some insects to enter. In addition to the beds, there were a chair, end tables, an area for hanging clothes, and a bathroom.

The bathroom featured a flush toilet, a sink with running water, and a cold water shower that had a roomy, one-piece fiberglass bottom. Linens, blankets, towels, and soap were provided. Bottled mineral water was placed in each guest's room, or filtered water was available on request.

There was no electricity in the guest cabins at La Selva. Lighting was provided by kerosene lanterns which were punctually lit by the employees each twilight. A generator did provide electricity to limited areas, primarily the dining and kitchen facilities where it aided in refrigeration and food preparation.

Guest cabins at La Selva. (Near Coca, Ecuador)

The two largest buildings at La Selva (immediately visible as one enters Lake Garzacocha) were the lounge and the dining hall. Each was roughly octagonal in shape and featured wall length windows that overlooked the lake and neighboring forest. Their location at the summit of a small hill also exposed them to an often welcome breeze. Chairs and tables were located

Main lodge building (R) and dining hall (L) at La Selva. (Near Coca, Ecuador)

throughout these buildings where guests could relax at their leisure. The food at La Selva was of high quality and served with pomp and style. Soda and drinks were available at reasonable rates from a well stocked bar.

Under construction at La Selva was a building to be used by biologists conducting research. This facility was to have living accommodations, running water, toilet facilities, a lab area, a library, and most probably electricity. The external structure of this field building was built during my visit in October 1988. (Note: Any intended use of this facility should be confirmed through Eric Schwartz prior to arrival at La Selva). I was informed in October 1989 the biological station had been completed and sleeps eight.

Trail System

La Selva had 3-4km of established trails on their property, as well as many native paths. A map indicating the major trails near the lodge can be obtained on arrival. Although there had been some slash and burn agriculture outside the La Selva grounds, the immediate area is basically undisturbed. Access to the rainforest of Yasuni National Park is simply accomplished by crossing the Napo River in a motorized canoe and hiking a short distance.

Costs

La Selva offers a variety of package tours to tourists at various prices, depending on the length of stay. Biological research is encouraged by the owners who may be willing to negotiate prices with scientists conducting legitimate research, in lieu of their providing partial guiding services. Biologists interested in participating in such an arrangement are advised to submit a brief proposal of the intended research and dates to Eric Schwartz, so that specific details can be worked out before arrival.

Comments

La Selva provides a very comfortable base from which to conduct long- or short-term tropical research. It is one of the few such facilities to be found in the Oriente of Ecuador, and is certainly the most luxurious. I was most impressed with the amount of visible animal life that could be seen on almost a daily basis. Several species of macaws and parrots fly noisily over the lake each afternoon as they go to roost. Yellow-rumped caciques nest just outside of the dining hall. Primitive hoatzins are commonly viewed no more than 15 meters from the lounge. In addition, two species of monkeys were seen during my brief visit from the vantage point of the lounge windows. At least eight species of primates have been reported at La Selva, as well as hundreds of species of birds. Although many faunistic and floristic studies of the immediate area have not yet been conducted due to the newness of the facility, indications are that the flora and fauna is especially rich. For example, the La Selva birdlist holds 450 species of birds and is still growing.

The construction of a field station facility is a tangible commitment of the La Selva owners to support and encourage long-term biological research at the lodge. Cooperative agreements between biologists and the La Selva management should mutually benefit all parties involved, and ultimately increase our knowledge of natural history and tropical biology of the area.

Jatun Sacha Biological Station

Contact

    Dr. David Neill (Director)

Address

    Estación Biologica Jatun Sacha
    Casilla 867, Sucursal 12
    Quito, Ecuador

        Or

    c/o Missouri Botanical Garden
    P.O. Box 299
    St. Louis, MO   63166-0299   USA
    Telephone: (314)-577-5100

ECUADOR

Description

The Jatun Sacha Biological Station is a rainforest preserve owned by the Ecuadorian non-profit organization Fundación Jatun Sacha. Jatun Sacha was established in 1985 with the aim of promoting research, eduacation, and conservation. Simple and rustic facilities have been constructed and are in the process of being expanded. Botanical and herpetological surveys conducted at Jatun Sacha have shown it to be one of the richest areas for species diversity in the world. The preserve is used by the School for Field Studies (see Chapter 3) as their South American field site. (The words *jatun sacha* mean 'old forest' in Quichua.)

Location

Jatun Sacha is located approximately 8km east of the small jungle town of Puerto Misahualli in the Oriente region of Ecuador. It is situated on the south side of the Rio Napo in Napo Province. The laboratory and living accomodations are located on a hill at about 400m elevation, a short distance south of the main road that leads from Puerto Napo to Campo Cocha. Jatun Sacha is situated between 0°-1° South latitude.

Logistics

The Jatun Sacha Biological Station may be reached in several ways. First, one can take a bus from Quito (or other cities) to Tena. From Tena you would take a bus to Puerto Napo, and then continue on it towards Campo Cocha. Jatun Sacha is located on the right (south) side of the road before reaching Campo Cocha, approximately 21km from the suspension bridge at Puerto Napo.

It is also possible to reach Jatun Sacha by motorized canoe from Coca. However, this entails traveling up the Rio Napo against the current which would take at least 8-10 hours, depending on the height of the river. One could then either get off and hike into Jatun Sacha, or continue by canoe all the way to Puerto Misahualli and then secure transportation by road. Since I did not take this route, I can not accurately describe what the river trip or hike from the Rio Napo would be like.

A third option is to rent a high clearance, 4-wheel drive vehicle and drive to Jatun Sacha. One possible route by car from Quito is described as follows. Leave Quito from the northeastern part of the city by getting on a road labeled as the 'Partidero a Tumbaco', which is located off of Avenida 6 de Diciembre. There will be a sign posted along this road before you leave Quito, indicating that the road goes in the direction of Tumbaco and Pifo. This road leads approximately east/southeast out of Quito, through the mountains. It is well paved for about 30 kilometers, where it turns into a bumpy, dirt road.

It is approximately 23km from Quito to Pifo. From Pifo you travel southeast along Rt. 28 until you reach Papallacta 39km away. It is then an additional 34km along Rt. 28 until you reach Baeza. At Baeza the road splits north and south, and is labeled Rt.45 on the map. The northern or left hand route continues on to Lago Agrio. You take the southern road towards Tena, which is 82km away. Several small towns such as Jondachi and Archidona will be passed along the way.

From Tena it takes approxiamtely 45 minutes to reach Jatun Sacha. There is a prominent statue of an Indian with a spear in his hand at the northern part of Tena (near the Hotel Auca). From this statue it is 28.7 km to Jatun Sacha. One continues past the statue along the main road and makes a left at the street marked Amazona. Another left at the next intersection will lead to a suspension bridge that spans the Misahualli River. Cross this bridge and stay on the main road until you see a carved bust in the median on your left. This point should be approximately 1.8km from the Indian statue. The road forks at this point and there is a CEPE gas station between the two ways. Take the left hand road which leads to Puerto Napo.

At Puerto Napo you will cross a suspension bridge that spans the Napo River. Immediately past the suspension bridge there will be a road to the left, which will lead to Jatun Sacha and eventually to Campo Cocha. From the suspension bridge at Puerto Napo, Jatun Sacha is approximately 21.1km away. About 5.2km from the Puerto Napo bridge, the road goes through the Rio Sindy. The road presents a rocky ford over this stream whose depth will depend upon recent rains (during my trip the water was about 10-12 cm deep coming and then 25-35 cm deep going).

The Jatun Sacha Biological Station is marked by a small sign on a high post along the right (south) side of the road at an entrance path. The lab and living facilities are located on top of a hill about a five minute walk from the road. Immediately across from the Jatun Sacha entrance path is a cut in the roadside wide enough to accomodate an automobile. Visitors with extensive luggage or equipment might want to park there to unload. Another 10-15 meters down the road on the right hand side is a larger parking area where vehicles will be less visible from the road.

Forest Type

The forest at Jatun Sacha is Tropical Wet Forest. It is on an unusually (for Amazonia) fertile clay soil.

Seasonality

Annual precipitation at Jatun Sacha is approximately 4,100mm. December and January are generally the driest months of the year.

Facilities

All buildings at Jatun Sacha were of wooden construction on raised decks, open-air, and covered with corrugated metal roofs. The laboratory was the main building, measuring approximately 7m x 12m and doubling as a food preparation and dining area. It contained two sinks with running water, gas run hot plates, dishes, glasses, and eating utensils. Work benches were available for dining or studying. A small library contained maps, information on the region, and reports of research conducted at Jatun Sacha. A large drying oven for plant specimens formed the major piece of equipment.

There were four sleeping houses with bunk beds that accommodated a total of 26 people. Two of these dormitories were open-air with meter-high siding, while the other was enclosed floor to ceiling and had sliding wooden shutters that fit in the windows. One had to supply one's own linens, blankets, towels, and toilet articles. Another building was the home of the

Sleeping quarters at the Jatun Sacha Biological Station. (Near Puerto Misahualli, Ecuador)

caretaker (Sr. Filadelfo Valencia). Bathing facilities were provided in the form of two covered, cement, cold-water shower stalls, with two adjacent sinks. These were located next to the laboratory. A four room, latrine-type outhouse was located at the edge of the clearing.

Laboratory and dining area of the Jatun Sacha Biological Station. (Near Puerto Misahualli, Ecuador)

A 110v electric generator and kerosene lanterns were available to provide electricity and illumination for visitors. One had to bring or supply one's own food at Jatun Sacha. There was a small store further down the main road towards Campo Cocha, or foodstuffs can also be purchased in Puerto Misahualli or Tena. An excellent restaurant also existed 2km further down the road from Jatun Sacha on the left hand side. It was part of the tourist facilities at Aliñahui Cabins (see pages 76-78). The proprietor of the restaurant prefers to be notified ahead of time if non-guests will be coming for dinner. One could boil water for drinking at Jatun Sacha or bring in bottled mineral water sold locally, or use the micropore water filter available.

Trail System

Approximately eight kilometers of unmarked trails exist throughout the preserve. A map of these trails is available on premises or by writing to Dr. Neill. The primary forest is adjacent to the station buildings. Jatun Sacha contains 300 hectares of land, almost 90% of which is undisturbed primary forest, but the adjacent properties near the road are being rapidly cleared.

Costs

The rate to stay at Jatun Sacha Biological Station and use their facilities was 2,500 sucres ($5) per night. A lower rate is charged to Ecuadorian nationals. One must also consider food costs. Bus transportation is inexpensive. However, if a rental car is used, a high vehicle with 4-wheel drive is suggested. These rented for approximately $40/day during my visit.

Comments

It is advisable that those people who intend to drive to Jatun Sacha from Quito do so during daylight hours. Large rocks are often in the roads, which tend to be narrow, curvy, bumpy, and mountainous during most of the trip. Landslides are not uncommon. In some of the higher altitudes the road becomes shrouded in clouds or mist, further decreasing visibility. If you misjudge your timing or are delayed, I would suggest overnighting at Baeza or Tena rather than continue on in the dark.

On a more positive note, the scenery along the roads in the mountains is spectacular. Approximately 44km out from Quito, between Pifo and Papallacta, is a boggy, moss-covered area with fascinating plant specimens and wildflowers. Further along in the area of Baeza, one can see numerous waterfalls cascading down from the forests on the side of the mountains.

The greatest attraction offered by Jatun Sacha and the surrounding area is the incredible biological richness of the region. This has been indicated in both faunistic and floristic studies. Jatun Sacha approaches the world record for the number of reptile and amphibian species at a single location (the current record is at Santa Cecilia, Ecuador, located approximately 140km NNE of Jatun Sacha). Botanical studies suggest that there may be more than 200 tree species per hectare at Jatun Sacha. Due to the recent establishment of the Jatun Sacha Biological Station and the fact that it is not yet well known among the scientific community, many groups of animals have yet to be sampled. However, species lists for birds and reptiles and amphibians are available.

The founders of Jatun Sacha are particularly interested in accommodating researchers wishing to conduct baseline biotic inventories of birds, small mammals, insects and other invertebrates, as well as other groups. A floristic study of Jatun Sacha and environs sponsored by the National Geographic Society is currently being carried out. Four, one-hectare, permanent tree plots have been established in different habitats. For those interested in visiting or working at Jatun Sacha, but preferring more luxurious facilities, it is suggested that the Aliñahui Cabins two kilometers away could be used as a base of operation.

Aliñahui Cabins

Contact

    Margarita Schenkel

Address

P.O. Box 5150 C.C.I.
Quito, Ecuador
Telephone: 479-831

The Aliñahui Cabins first opened in December 1987. They provide lodging and meals to tourists and visitors to the Oriente region. Located approximately 10km east of Puerto Misahualli, the Aliñahui facilities are only two kilometers further than Jatun Sacha, along the road between Puerto Napo and Campo Cocha (see previous section, page 70). Aliñahui is approximately 23.1km from the suspension bridge at Puerto Napo.

The entrance road to Aliñahui was marked with a small, raised, red wooden hut on the north side of the road. One turns left here and immediately comes to a white gate (which the occupant of the red hut will probably come out and open, although it is unlocked). One proceeds through the gate and follows the entrance road for approximately a kilometer. There will be pasture on either side and then a cleared area for parking at the end of the road. The resataurant and cabins are located a short walk beyond.

The buildings of Aliñahui were situated on a bluff overlooking the Napo River (hence the name *aliñahui* which means 'nice view' in Quichua). There were a total of six guest cabins that were capable of accommodating 24 people. Each cabin had two screened bedrooms containing two beds each. Linen, towels, blankets, and soap were provided, while solar-powered lights furnished illumination.

The sleeping quarters of each cabin were elevated on stilts approximately four meters above a cement slab. A stairway provided entrance to the rooms with a comfortable open-air area with table and chairs in between. The lower area beneath the cabin on the slab also had table and chairs, as well as hammocks and hanging plants. There was one bathroom for each cabin, located on the slab immediately adjacent to the stairway. Bathrooms were screened and spacious, and featured a sink with running water, cold water shower, and a flush toilet. There was a light in each bathroom.

Elevated guest cabin with detatched groundlevel bath at the Cabañas Aliñahui. (Near Puerto Misahualli, Ecuador)

The food at Aliñahui was simple, but excellent. Soda, beer, and liquor was available at a reasonable price, while filtered water was available at no charge. Three meals a day were included with the price of the rooms, which during my stay was $10/day (5,000 sucres/day).

I strongly recommend the Aliñahui Cabins to anyone visiting this area of Ecuador. The multi-lingual owners maintain a clean and efficient operation, at an extremely reasonable price. They are very knowledgeable about the area and can provide much useful information. Part of the Aliñahui land is held in rainforest, but I did not have the opportunity to visit this area during my stay.

Maquipucuna Tropical Reserve

Contacts

    Rodrigo Ontaneda
    Casilla 167-12
    Quito, Ecuador
    Telephone: 23-38-71 or 23-61-66

The Maquipucuna Tropical Reserve (MTR) consists of 2,500 hectares of undisturbed premontane rainforest situated on the western slopes of the Andes. It is located approximately 45 minutes northwest of Quito, past the town of Calacali. Elevations at the MTR range from 1,200 to 2,800 meters. Annual precipitation varies from 1,000 to 4,000 millimeters. The MTR was recently established by the Fundación Maquipucuna with aid from the Nature Conservancy. The Fundación Maquipucuna was founded in 1987 and is a conservation-oriented, non-governmental, non-profit organization.

The field station associated with the MTR opened during August 1989. At present, it consists of a large cabin located at approximately 1,200m elevation approximately 1km from the actual reserve. The cabin can accommodate approximately 15 people and has a kitchen, flush toilet, and showers. There is no electricity or generator at the moment. The nearby town of Nanagal is 20-30 minutes by car and does not offer much in the way of supplies. Therefore, researchers should plan on bringing all needed items with them. Although a few trails are available in the reserve, there is currently no established trail network or system. The Fundación Maquipucuna intends to establish a network of tropical forest reserves throughout Ecuador. Additional information on the Fundación Maquipucuna can be obtained at the above addresses.

Rio Guajalito Reserve

Contact

> Dr. Jaime Jaramillo
> Universidad Católica del Ecuador
> Apartado 2184
> Quito, Ecuador

The Rio Guajalito Reserve is located off the old road from Quito to Santo Domingo at an elevation of approximately 1,800 meters. It is composed mostly of primary forest, but with some land in pasture and secondary forest as well. More detailed information can be obtained from botanist Dr. Jaime Jaramillo at the above address.

Quito

Mariscal Sucre Airport is a modern facility located in the northwestern portion of Quito. I had no problems with luggage on either the Eastern or TAME flights, and passing through customs was a formality. One should be forewarned that the exit tax for leaving Quito on any international flight is $25, to be paid in cash (either dollars or sucres) before you are allowed to enter the gate.

I stayed at the Hotel Embassy, located on Calle Presidente Wilson near the corner of Avenida 6 de Diciembre. Taxis were easily found at the airport and the 20 minute ride to the hotel cost between $1-$2 (500-1,000 sucres), depending on the amount of luggage. The Embassy is a modern hotel with 60 rooms on several floors, as well as adjacent detatched rooms. There is a bar and restaurant on the ground floor of the hotel where food was very inexpensive and of good quality.

A double room with two beds at the Hotel Embassy cost approximately $12/night (6,000 sucres/night), including taxes. The hotel is often busy with foreign tour groups on their way to the Galapagos Islands, so that reservations should be booked in advance. Within walking distance of the Hotel Embassy are the Bureau of Tourism, the Universidad Católica, the Casa de las Culturas, money changing houses, souvenier shops, and many fine restaurants.

Quito is located at an altitude of approximately 9,300 feet. This may present problems to some people in the form of fatigue, shortness of breath, and/or headaches. It is best not to exert too much physically the first day or two in Quito so that your body has time to adapt. Taxis are extremely inexpensive and can usually be found without difficulty.

The car rental agencies are found adjacent to the airport. Although some companies maintain branch offices within the city, a better selection of vehicles can usually be obtained at the main offices near the airport. I used the company Ecuacar, which furnished vehicles that functioned generally well. The cost of renting a 4-wheel drive Chevrolet Trooper came out to approximately $40/day. A vehicle with high clearance is strongly recommended if any driving outside of Quito is planned. Gas is very inexpensive and cost about 40 cents per gallon at the time of my trip.

Anyone intending to rent a car should be ready to sign away their life. There is an automatic deductible of several hundred dollars on almost every rental contract, in the case of an accident. Often the insurance does not cover windshields, windows, bumpers, and lights. If the car is stolen, the renter may have to continue paying the daily rental charges on it for as long as three months afterward! Ecuadorian cars have special locks built in near the stick shift to prevent theft. It functions by first placing the car in reverse and then sliding a thick, U-shaped metal bar (similar to a bicycle lock) around the stick and into a permanently mounted receptacle. This lock can then only be opened with a key. A portion of the car renting process includes examining the auto and marking down any broken, scratched, or malfunctioning items. Take this process seriously and do a thorough job, so as to avoid any problems when you finally turn the car in.

Tena

Tena is the capital city of Napo Province, located at an elevation of approximately 600 meters. The original vegetation ranges from Tropical Wet Forest to Premontane Wet Forest. My observations of Tena are restricted primarily to the Hotel Auca. One reaches the Hotel Auca when coming from the direction of Baeza by following the main road through town until the statue of the Indian with a spear in his hand is reached. This statue stands in the middle of a traffic circle. One goes three fourths of the way around the circle and then turns to the right. Approximately 0.5km down this road, one will see a sign on the right hand side marking the entrance drive of the Hotel Auca. The entrance road leads 0.2km to the hotel, which is on a bluff overlooking the Misahualli River.

The Hotel Auca may appear shabby by North American standards, but it is perfectly tolerable for a few nights. There are eight small cement block buildings with two rooms each, and three beds to a room. Each room had carpeting, an electric fan, a closet, and a bathroom with sink, flush toilet, and cold water shower. Electricity typically ran from 6:30am - 12:00 midnight, providing 110V service in the room.

The hotel has a dining room and restaurant, as well as a bar. There are no menus in the restaurant, and one is served whatever dish happens to be prepared for that particular meal. The food was simple, but adequate and quite inexpensive. Room rates were also inexpensive at approximately $6 per night.

Comments On Ecuador

Ecuador is a country of great geographical diversity and biological richness confined within a relatively small area. Petroleum exploration and development has resulted in the construction of serviceable roads into the sparsely populated and heavily forested Oriente region, but also threatens the pristine quality of protected areas such as Yasuni National Park. While tourism has existed in Ecuador for a long time due to the Galapagos Islands, the development of natural history tours into the rainforests of the eastern regions are just starting to be promoted.

Ecuador is one of the most inexpensive countries in Latin America to visit or work in. A variety of facilities are accessible at various locations in relatively undisturbed areas for use by tourists or researchers. The fauna and flora is incredibly rich and the scenery spectacular. Ecuador can easily provide something for everyone.

Books

There are relatively few books available that treat the flora and fauna of mainland Ecuador, most being devoted to the Galapagos Islands. The most complete bookstore in Quito is Libri Mundi located at 851 Calle Juan Leon Mera, between Calle Carrion and Calle G. Veintimilla. Libri Mundi also maintains a branch office in the Hotel Colon. Other bookshops can be found in the shopping district along Avenida Amazonas.

GALAPAGOS: WORLD'S END
Beebe, William
1924. G.P. Putnam's Sons

ECUADOR

FLORA OF THE RIO PALENQUE SCIENCE CENTER
Dodson, C.H. and A. Gentry
Selbyana 1978

FLORA OF JAUNECHE
Dodson, C.H., A. Gentry and F.M. Valverde
Banco Nacional de Ecuador and Selbyana 1985
(Obtainable from the Missouri Botanical Gardens Publication Office)

MEDIO AMBIENTE Y DESARROLLO EN EL ECUADOR
Encalada Reyes, Marco A.
Salvat Editores Ecuatoriana S.A.
Fundación Natura

TROPICAL NATURE: LIFE AND DEATH IN THE RAINFORESTS OF CENTRAL AND SOUTH AMERICA
Forsyth, Adrian and Ken Miyata
1987. Charles Scribner's Sons

ADVENTURING IN THE ANDES: THE SIERRA CLUB TRAVEL GUIDE TO ECUADOR, PERU, BOLIVIA, THE AMAZON BASIN, AND THE GALAPAGOS ISLANDS
Frazier, Charles with Donald Secreast
1985. Sierra Club Books

RÍO NAPO -- REALIDAD AMAZONICA ECUATORIANA
Gonzales, Angel and Juan Santos Ortiz De Villalba
1985. Ediciones CICAME

GALAPAGOS: A NATURAL HISTORY GUIDE
Jackson, Michael H.
1985. University of Calgary Press

ECUADOR: ISLAND OF THE ANDES
Kling, Kevin and Nadia Christensen
1988. Thames and Hudson

ECUADOR AND THE GALAPAGOS ISLANDS: A TRAVEL SURVIVAL KIT
Rachowiecki, Rob
1986. Lonely Planet Publications

TWO WHEELS AND A TAXI: A SLIGHTLY DAFT ADVENTURE IN THE ANDES
Urrutia, Virginia
1987. Mountaineers

Maps

Various maps of Ecuador can sometimes be purchased in bookstores or from vendors and shops at the airport. However, the most complete selection of recent and accurate maps is available from the Instituto Geográfico Militar, located on top of a high hill at Calle G. Paz y Mino and Calle Seniergues. Domestic sources of Latin American maps are listed on pages 290 and 291.

Tourist Information Sources/Travel Agencies

South American Explorer's Club
Apartado 21-431
Eloy Alfaro
Quito, Ecuador
Telephone: 55-60-76
Street Address: Toledo 1254
  La Floresta, Quito

South American Explorer's Club
1510 York Street
Denver, CO 80206 USA
Telephone: (303)-320-0388

Dirección Nacional de Turismo
Reina Victoria 514 and Juan Leon Mera
Quito, Ecuador
Telephone: 239-044 or 527-002

ECUADOR

Asociación Ecuatoriana de Agencias de
   Viajes y Turismo--ASECUT
Amazonas 657 y Ramirez Davalos
Casilla 1210
Quito, Ecuador
Telephone: 529-253

Empresa Ecuatoriana de Aviación - EEA
Edificio Rocafuerte
Avda. Jorge Washington 718
Apartado 505
Quito, Ecuador

Adventure Associates
13150 Coit Road
Suite 110
Dallas, Texas 75240 USA
Telephone: (214)-907-0414

Ecuadorian Castle and Mountain Tours, Inc.
Ingraham Building
25 S.E. 2nd Avenue, Suite 540
Miami, Florida 33131 USA
Telephone: (305)-577-0559

Conservation Organizations

Asociación Ecuatoriana Para La Conservación De La Naturaleza
Las Casas 18-58
Quito, Ecuador

Instituto Ecuatoriano de Ciencias Naturales
Apartado 408
Quito, Ecuador
Telephone: 215-497

Fundación Natura
Avenida 6 de Diciembre y El Comercio
Casilla 253
Quito, Ecuador

Fundación Ornitológica Del Ecuador
Casilla 9068 S-7
Quito, Ecuador

Scientific Organizations/Institutions

Charles Darwin Research Station
Puerto Ayora
Santa Cruz
Galapagos Islands
Ecuador

Casa de la Cultura Ecuatoriana 'Benjamin Carrion'
Apartado 67
Avenida 6 de Diciembre 794
Quito, Ecuador

Instituto Interamericano Agricultural Experimental
Conocoto
Linea 63
Quito, Ecuador

Instituto Nacional de Investigaciones Agropecuarios
San Javier 295 y Orellana
Apartado 2600
Quito, Ecuador

Instituto Nacional de Higiene y Medicina Tropical
　'Leopoldo Izquieta Perez'
Apartado 3961
Guayaquil, Ecuador

ECUADOR

Institut Francais de Recherche Scientifique por
  le Developpement en Cooperation (ORSTOM)
Apartado 6596, CCI
Quito, Ecuador
Telephone: 544-122

Universidad Central Del Ecuador
Avenida America y A. Perez Guerrero
Apartado 3291
Quito, Ecuador
Telephone: 524-714

Pontificia Universidad Católica Del Ecuador
Avenida 12 de Octubre 1076 y Carrion
Apartado 2184
Quito, Ecuador
Telephone: 529-240 or 529-280

CHAPTER 1C   FRENCH GUIANA

General Information

Area: 90,000 sq km (34,750 sq miles)

Population: 100,000 (approximate)

Capital: Cayenne (population: approximately 65,000)

Language: French

Currency: French francs (1 French franc = 100 centimes)

    Coins: 1, 5, 10, 20 and 50 centimes; 1, 2, 5 and 10 francs.
    Notes: 20, 50, 100, 200 and 500 francs.
    Exchange Rate: U.S. $1 = 5.5 francs (June 1988)
                    U.S. $1 = 5.7 francs (March 1990)

National
Airlines: Air France, Air Guyane

Telephone   French Guiana: 594
Code: No special city codes are needed.

French Guiana

1 – Placer Tresor   2 – ORSTOM Facilities   3 – Lassort Lodge

FRENCH GUIANA

Introduction

French Guiana (which is called Guyane in French) is a French overseas department as are Guadeloupe and Martinique. French is the official language, and few of the inhabitants speak anything other than French and Creole. Approximately 90% of French Guiana is undisturbed tropical rainforest. Almost half of the department's population is centered in Cayenne. The majority of the rest of the people are concentrated in the several large cities along the coast and river systems, the largest of which has a population of about 8,000. The department's interior is sparsely populated and attracts a small number of tourists and hunters.

French Guiana is located in the northeastern portion of the South American continent. It is bordered to the west by Surinam and to the south and east by Brazil. Its northern boundary is its coastline along the Atlantic Ocean. The terrain is composed of heavily forested coastal plains in the northern half of the department, giving rise to hills and the Tumuc-Humac mountain range in the southern portion. French Guiana is approximately the size of the state of Indiana.

The near pristine condition of the vast majority of rainforest in French Guiana offers a unique opportunity to both tourist and biologist. Limited scientific study has been conducted, primarily through the efforts of ORSTOM (the French overseas research organization) and The New York Botanical Garden. Unfortunately, French Guiana has few restrictions on the collecting/killing and removal of plants and animals, which makes its wildlife particularly vulnerable to exploitation.

My evaluation of rainforest sites in French Guiana took place in June 1988. During this time I visited sites deep within the interior (Saül), within a day's drive of the capital (Placer Trésor), and on the Maroni River along the western border with Surinam (Lassort Lodge).

Placer Trésor

Contact

    Jacques Riché

Address

    97311 Roura
    France

Description

    Placer Trésor consists of several hammock houses and a small restaurant situated in a clearing in the middle of the rainforest. It is used primarily by hunters and tourists who wish to take advantage of the owner's vast knowledge of the forest and of French Guiana. It is operated by the Riché family who live on site in the dry season, and on the weekends during the rainy season.

Location

    Placer Trésor is approximately a 25km drive from the town of Roura, which is located 23.5 km south of Cayenne on the east bank of the Oyac River. Placer Trésor is situated between 4°-5° North latitude.

Logistics

    To reach Placer Trésor by car, one drives south from Cayenne in the direction of Rochambeau Airport (Aérodrome). Before actually reaching the airport, one will see road signs indicating Roura. Following the signs to Roura, one will come to the Stoupan Ferry, which transports cars (approximately eight at a time) free of charge across the Mahury River. The ferry operates once an hour from 6:00 am - 6:00 pm, with the exception of 1:00 - 3:00 pm. It should take no longer than 20-30 minutes to reach the Stoupan Ferry from Cayenne.

After crossing the Mahury River, the town of Roura is 5-10 minutes away by car. Mr. Riché and his family have a house in Roura, located behind the blue schoolhouse that you must pass as you enter the town on the road from the ferry. One makes the first right after the schoolhouse, and then the next right to reach the Riché residence.

To reach Placer Trésor from Roura, you must make the first left past the schoolhouse from the road leading into the town and then bear left until you come to a T-intersection. At this intersection, a sign indicates 'Gabrielle Restaurant' to the left and 'Crique Gabrielle' and 'Kaw' to the right. One turns right and follows the road for approximately 3km until a fork is reached. Here, a sign will indicate 'Crique Gabrielle' to the left and 'Fourgassié' to the right. Bear to the right and follow the road for approximately 9.5km until another fork is reached. Here, a sign indicates 'Auberge de Brousse des Cascades' to the right, and there may or may not be a legible sign indicating 'Camp Caiman' to the left. One bears to the left and continues for approximately 5.4km until a side road appears on the right with several signs posted indicating the tariffs and facilities available at Placer Trésor. This is the access road to Placer Trésor and it continues for approximately 4km. At its end is a small grassy area in a clearing of the forest, which contains the actual facilities.

Forest Type

The slopes of the Kaw Mountains where Placer Trésor is located are covered by Tropical Moist Forest. The Kaw Mountains are a series of lateritic table mountains with their plateau at an altitude of approximately 400m.

Seasonality

The rainy season lasts from approximately January to June with a peak in May and a break in March. The dry season lasts from approximately August to December.

Facilities

The facilities of Placer Trésor included an open-air restaurant, a screened cabin for 8-10 hammocks, and a cabin

Hammock house and sleeping quarters at Placer Trésor. (Near Roura, French Guiana)

under construction for 2-4 beds. There were two flush toilets, two showers, and a sink that used river water, adjacent to the hammock cabin. All structures were covered with corrugated tin roofs and were provided with light bulbs that derive their electricity from an on-site generator. The hammocks were provided, and one could either eat at the restaurant or bring in one's own food.

Trail System

There were a number of unmarked trails that lead out from the central area where the buildings were located. It is advisable to learn the trail system under Mr. Riché's guidance, to prevent becoming lost.

Costs

I was charged approximately 35 francs ($6.50) per night by Mr. Riché solely for the use of a hammock and the other facilities, combined with access to the grounds. I brought in my own food (bread and canned goods) to keep costs down. Car rental cost me 195 francs ($36) per day and was arranged in Cayenne. If one is planning to spend several days or more at Placer Trésor, I would suggest arranging to have Mr. Riché pick you up in Cayenne and drive you to his facility.

Comments

For visitors intending to drive to Placer Trésor, make sure your car has a spare tire, as the road from Roura is extremely rocky and almost completely devoid of habitation. The 4km access road to Placer Trésor was even worse, being deeply rutted and very muddy in spots. There are also no gas stations in Roura, so it is advisable to fill your tank in Cayenne.

Mr. Riché has his own van which he sometimes uses to transport his guests from Roura to Placer Trésor. It is advisable to confirm such plans ahead of time. A taxi from Cayenne to Roura, or all the way to Placer Trésor, may be advisable for visitors who are hesitant about driving on bumpy, mountain roads.

Placer Trésor is located at an area that used to support gold mining approximately 50 years ago. Some of the mining machinery is still in evidence, but the forest is in a generally undisturbed condition. The combination of restaurant, modern toilet and bath, and screened-in sleeping facilities offers an unusual degree of comfort for visitors to the rainforest in French Guiana. Simple foodstuffs can be purchased in Roura, which also has a small, but excellent six room hotel called the Oyak.

The Hotel Oyak is recommended to visitors who arrive by car in Roura at or near dusk, as the trip to Placer Trésor from Roura can take up to an hour due to the condition of the roads and the need to drive slow. The Oyak is located on a scenic branch of the Comté River and has its own small restaurant. It can be reached by making the first right past the schoolhouse upon entering Roura, and then the first left. It is on the right side of the road along side of the church. Reservations should be made before arrival to assure that the hotel will be open(Phone: 30-41-93).

Saül

Contact

(See Comments section on page 98.)

Description

Saül is an old gold mining village where some of the inhabitants still pan for gold. Though the tourist map of French Guiana lists the population as 119, approximately 40 is more accurate. Relatively undisturbed forest can be found 15 minutes away from the village, but most larger animals have been hunted out nearby.

Location

Saül is located in the center of French Guiana, approximately 300km southwest of Cayenne. It is at an elevation of approximately 200 meters and situated between 3°-4° North latitude.

Logistics

Saül is most easily reached by airplane. Air Guyane maintains regularly scheduled flights to Saül from Cayenne. Flying time is approximately 45 minutes. Upon arrival, there is a 20-30 minute walk from the airstrip to the village. Sometimes a tractor-driven wagon is available to transport the luggage, but this is not always the case.

Forest Type

Tropical Moist Forest surrounds the area and settlement of Saül. Within a day's walking distance elevations range from 200 to 750 meters.

Seasonality

There is a well defined dry season that lasts from August through November. The wet season occurs from December through July. More detailed information about the climate in Saül can be obtained by consulting volume 44 of the Memoirs of the New York Botanical Garden. The title of this volume is *The Lecythidaceae of a Lowland Neotropical Forest: La Fumée Mountain, French Guiana*, by Scott A. Mori and collaborators.

Facilities

ORSTOM (see page 106) maintains basic facilities at Saül for use by its scientific staff during field research periods. The ORSTOM building contains three unscreened rooms large enough to accomodate two hammocks each. The rooms have shuttered windows for ventilation. Adjacent to these rooms is a combination lab room/dining room of approximately 4 x 5 meters. Beyond the lab/dining room, separated by a partial wall, is the cooking area, sink, and shower. There was a gas operated camping stove available during my visit. Toilet facilities consisted of a detatched outhouse behind the building.

Water for the sink is brought down by gravity from a holding tank connected to a resevoir. I drank this water, unfiltered, with no ill effects. However, one time I did find tiny tadpoles in my glass. Breakfast and dinner used to be available through a village resident named Michel Modde. However, Mr. Modde passed away in June 1988 and visitors should plan to bring in their own supplies unless otherwise advised. Very limited foodstuffs were available at a small store in Saül, and one inhabitant named Annette operated a restaurant from her home. The meals provided there were quite satisfactory.

The status of tourism in Saül is presently in a confused state. Mr. Modde formerly operated cabins which he rented to tourists. It is not known however, if these structures will be continued to be maintained in his absence. I have been informed that hammock space could be rented in a house owned by M. Gregory (nicknamed Popo) for 15 francs/person/day. Space was also available in a hammock shelter operated by Guy Bourdot for 20 francs/person/day.

Trail System

There are nearly 100km of marked and well-maintained trails in the vicinity of Saül. Trail markers consist of a band of color painted on trees along a given trail. A map outlining these trails has previously been available through ORSTOM.

ORSTOM factilities at Saül. (French Guiana)

Costs

During my visit to Saül, I stayed in a room owned by the church and located across from the ORSTOM house. The cost was approximately 12 francs ($2) per night. I did the majority of my own food preparation with foodstuffs I had purchased in Cayenne (note: all imported food items are very expensive). The one dinner I had at Mr. Modde's house cost 60 francs ($11), which was also the cost of the average dinner at the house of Annette. The round trip airfare from Cayenne to Saül on Air Guyane cost 625 francs ($113), and had increased to 660 francs by late 1989.

Comments

Tourists or biologists interested in visiting Saül should try to obtain information from the various organizations listed later on in this section. ORSTOM and The New York Botanical Garden (see page 106) will probably be the most helpful. It is strongly advised to write in French when soliciting information from any organization located in French Guiana or mainland France.

Lassort Lodge

Contact

    Madame or Jean Robert Lassort

Address

    B.P. 55
    St. Laurent du Maroni
    France
    Phone: 31-49-45 or 31-61-90

Description

    The Lassort Lodge is a tourist facility that can provide both lodging and meals to visitors. It is owned and operated by the Lassort family. It consists of several hammock houses and a main lodge building which is used for food preparation and dining.

Location

    The facility operated by the Lassorts is located near the town of Maripasoula, situated near French Guiana's western border with Surinam. The specific area is called Saut Sonelle and is located on a curve of the Inini River. Maripasoula is approximately 95km due west of Saül and 235km southwest of Cayenne. The Lassort Lodge is situated between 3°-4° North latitude.

Logistics

    The Lassort Lodge is located approximately 10km northeast of Maripasoula. The town of Maripasoula contains approximately a thousand inhabitants and can be reached most easily by air, through the regularly scheduled flights of Air Guyane. As a visitor to the lodge, I took a 10 minute taxi ride from the airstrip to the *gendarmerie*, where I had been told to wait. I was then met by Jean Robert Lassort, who showed me to a scenic restaurant in Maripasoula where I could have lunch and wait to be picked up two hours later.

Upon Mr. Lassort's return, we walked back to the *gendarmerie* to pick up our luggage, and proceeded to the river bank a short distance away. Our gear was loaded in a long, narrow, open, motorized canoe called a *pirogue*. A non-stop trip from Maripasoula to Saut Sonelle takes about one hour, while the return takes about 30 minutes. Our trip out to the lodge was interrupted by a stop at an Indian village, whose inhabitants exist in a mixture of their traditional and western cultures.

Forest Type

The area around the Lassort Lodge is covered with Tropical Moist Forest, with elevations of approximately 90 to 100 meters.

Seasonality

The rainy season in French Guiana lasts from January to June, peaking in May and with a break in March. The dry season lasts approximately from August to December.

Facilities

The main lodge building and restaurant was an open-air, wooden structure covered with a corrugated metal roof. The greater portion of it (approximately 7m x 15m) contained a large U-shaped table (capable of seating about 30 people) on a gravel floor. The remaining part contained the cooking area and a small lounge overlooking the Inini River. This main building was equipped with electricity, provided by an outdoor generator.

The sleeping habitations were open-air, A-frame structures with half walls built on a raised wooden platform. Hammocks with mosquito nets and sleeping bags were provided for the guests. During my visit, two such hammock houses with thatched roofs and containing two hammocks each were present, as well as one larger one with a shingled roof containing five hammocks. Two other less rustic wooden habitations also appeared near completion. There was no electricity in the hammock houses. Toilet facilities consisted of one semi-enclosed outhouse, with a hole in a wooden platform over a pit. Bathing was done in the river.

Trail System

There did not appear to be a well developed trail system, as only two unmarked trails were easily accessible. These led out from either side of the lodge and paralleled the river. Each was narrow, and after 1-2 kilometers was interrupted by high water (a condition which might only prevail during the rainy season).

Costs

The rates for staying at the Lassort Lodge were dependent upon whether or not one ate meals prepared by the owners. To save money, I brought canned goods and bread with me. The cost then for just sleeping in one of the hammock houses and being picked up and transferred from Maripasoula to the lodge (via *pirogue*) was not high. Detailed information about rates can be obtained by writing to the Lassorts at the address provided earlier (Note: it is strongly suggested that you write in French). The cost for a round trip ticket from Cayenne to Maripasoula on Air Guyane was 750 francs ($136).

Comments

As in the rest of French Guiana, unless one speaks some French, communication is extremely difficult. I was hosted by Jean Robert Lassort who spoke some English, but not enough for any meaningful conversation. This eliminated the chance of learning much about the area from those most familiar with it. Excursions were explained in French.

Photographs in a scrapbook of various animals, birds, snakes, etc. killed in the area of the lodge, as well as pelts used as decorations, supported the feeling of 'macho hunter' that I received from inhabitants and visitors throughout French Guiana. Aside from ORSTOM scientists, no one seemed much concerned with conservation.

Hammock house and sleeping quarters at the Lassort Lodge. (Near Maripasoula, French Guiana)

A French guide that I met in Saül informed me that Madame Lassort intended to operate a combination hiking/boating excursion through the forest. Groups originating in Saül would hike to Dorlin (approximately 67km and 4-5 days travel away). From Dorlin, there would be another 3-5 days of traveling by *pirogue*, depending upon the water level of the river. I have more recently been informed that Guy Bourdot (see page 97) brings tourists to the Lassort facility from Saül.

Cayenne

My stay in Cayenne was made much more pleasant by residing at the Hotel Neptima, where the proprietors operated a clean hotel of 15 rooms. Prices per room ranged from 90-175 francs ($16-$32) per night. The Neptima is located near the corner of Rue Félix Eboué and Rue Christophe Colomb. It is convenient to grocery stores, bakeries, bookstores, and shops, as well as the main plaza (where the offices of Air France and Air Guyane are located). Overseas phone calls could be placed from the hotel lobby. An excellent breakfast was available to hotel guests at a cost of 20 francs ($3.60).

FRENCH GUIANA

Comments On French Guiana

French Guiana is one of the most expensive areas in South America for travelers. This ranges from airfare, to food, to hotels, to gas. One should arrive with French money already in hand, as other forms of currency will not usually be accepted. (A 10-franc coin is indispensable for using the baggage carts at Rochambeau Airport in Cayenne.)

Travelers should visit the tourism office (located at the botanical garden) to pick up free literature about excursions and services available in French Guiana, as well as a map of Cayenne. The *Mini Guide GUYANE* (written in English) was an especially helpful booklet that listed hotels, restaurants, tour operators, and much more additional information. In addition to the rainforest, visitors may find the following of interest: the space center at Kourou, Devil's Island (the scene of the book *Papillon*), Mana (where giant sea turtles spawn from April to July), and Cacao (a Laotian village whose inhabitants raise shrimp in large basins). It is highly advisable that at least one person in the group speak French in order to facilitate communication.

Books

Most of the following books can be purchased in bookstores and shops in Cayenne.

SERPENTS DE GUYANE
Chippaux, Jean Philippe; Léon Sanite, Daniel Heuclin
SEPANGUY/Nature Guyanaise Series (Cayenne)

FAUNE DE GUYANE
Delabergerie, Guy
1985. Editions G. Delabergerie (Cayenne)

FLORE DE GUYANE
Delabergerie, Guy
1985. Editions G. Delabergerie (Cayenne)

LES TORTUES DE GUYANE FRANCAISE
Frétey, Jacques
1987. Nature Guyanaise Series

FLORE ET VEGETATION - LA DOCUMENTATION GUYANAISE
de Granville, Jean-Jacques
1986. Saga

PHARMACOPEES TRADITIONELLES EN GUYANE CREOLES, PALIKUR, WAYAPI
Grenand, Pierre; Christian Moretti, Henri Jacquemin
1987. Editions de l'ORSTOM (Paris)

THE LECYTHIDACEAE OF A LOWLAND NEOTROPICAL FOREST; LA FUMEE MOUNTAIN, FRENCH GUIANA
Mori, Scott A. and Collaborators
1987. Memoirs of the New York Botanical Garden Volume 44

ORCHIDEES DE GUYANE
Pawilowski, Claude
SEPANGUY/Nature Guyanaise Series (Cayenne)

LE LITTORAL GUYANAIS - FRAGILITE DE L'ENVIRONMENT
1986. SEPANGUY/SEPANRIT. Nature Guyanaise Series (Cayenne)

Maps

A tourist map of Guyane and its roads *(Carte Touristique et Routiére/* 1:500,000) can be purchased at most bookstores for approximately 30 francs. A more specific map (# 516) of the coastal area from Cayenne to Kourou is also available at the same price. One may be able to purchase more detailed topographical maps from the Institut Géographique National in Paris. Maps of Cayenne can be obtained for free at the Tourism Office in Cayenne, at the Botanical Garden.

FRENCH GUIANA

Tourist Information Sources

Agence Régionale de Développement du Tourisme et des Loisirs
Pavillon du tourisme - Jardin Botanique
P.O. Box 801 - 97338 - CAYENNE Cedex
France
Telephone: 30-09-00
Telex: ARTOUR 910356 FG

Déleégation Régionale au Tourisme
10, rue L.-Heder
97307 CAYENNE Cedex
France
Telephone: 31-84-91
Telex: PREFGU 910532

French Government Tourist Office
610 Fifth Avenue
New York, N.Y. 10020 USA
Telephone: (212)-757-1125

Conservation Organizations

SEPANGUY
Société d'Etude, de Protection, d'Aménagement de la Nature en Guyane
Dr. Sanite
Avenue Pasteur
Bureau Veterinaire Departemental
Cayenne
France

SEPANRIT
Société Pour l'Etude, la Protection et l'Aménagement de la Nature Dans Les Regions Inter-Tropicales
Dr. Sanite
Avenue Pasteur
Bureau Veterinaire Departemental
Cayenne
France

Scientific Organizations/Institutions

ORSTOM
Institut Francais de Recherche Scientifique pour le Déve loppement en Coopération
Centre de Cayenne
BP 165
97323 CAYENNE Cedex
France
Telephone: 30-27-85   Telex: 910608

Note: ORSTOM has a large facility in Cayenne, including a herbarium and excellent botany staff. The ORSTOM scientists have made numerous field trips throughout French Guiana and can be extremely helpful to other biologists and colleagues working in the department of French Guiana.

The New York Botanical Garden
Bronx, NY 10458-5126 USA
Telephone: (212)-220-8700

INRA
Institut National de la Recherche Agronomique
BP 351
97310 Kourou
France
Telephone: (594)-32 13-10   Telex: 910644 FG

IRAT
Institut de Recherches Agronomiques Tropicales
BP 60
97322 CAYENNE Cedex
France
Telephone: 31-10-96

Institut Pasteur de la Guyane Francaise
97306 CAYENNE Cedex
France
Telephone: 30-17-66   Telex: INPASST 910573 FG

CHAPTER 1D   VENEZUELA

General Information

Area: 912,050 sq km (352,143 sq. miles)

Population: 19,246,000

Capital: Caracas (population: 4,000,000)

Language: Spanish

Currency: Bolívar (1 Bolívar = 100 centimos)
Coins: 5, 12 ½, 25, and 50 centimos
1, 2, and 5 Bolívares
Notes: 5, 10, 20, 50, and 100 Bolívares
Exchange Rate: U.S. $1 = 38 Bolívares (June 1989)
U.S. $1 = 43 Bolívares (March 1990)

National
Airlines: VIASA, AVENSA, Aeropostal

Telephone Venezuela   58
Codes: Caracas   2
           Maracay   43
           Merida   74
           Puerto Ayacucho 48

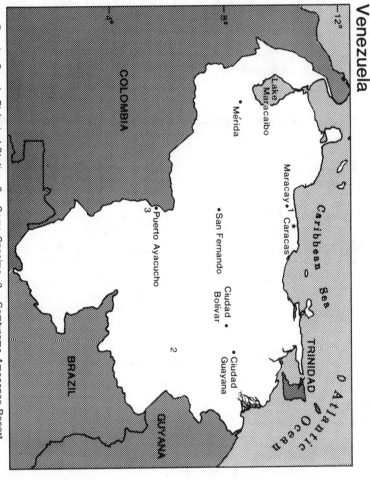

VENEZUELA

Introduction

Venezuela is located on the northern coast of South America in an approximately central position. It borders Guyana to the east, Brazil to the south and southeast and Colombia to the west and southwest. Venezuela has more than 2,800 km of coastline along the Caribbean Sea and for a small portion along the Atlantic Ocean on its northern boundary. It is a large country, roughly the size of Texas and Oklahoma combined.

Venezuela is divided into four distinct geographical regions. The Orinoco River, flowing from southwest to northeast, effectively bisects the country. To the south and east of the Orinoco River are the Guyana Highlands. This is a sparsely populated region of rolling plains and high plateaus. In this area is Angel Falls, many of Venezuela's flat-topped mountains called *tepuis*, and Canaima National Park. In the Federal Territory of Amazonas in southernmost Venezuela is a lowland area drained by the Rio Negro, Rio Casiquiare, Rio Ventuare, and the northwesterly flowing portion of the Rio Orinoco.

The three other major geographical regions occur north and west of the main Orinoco River. They begin with the Andes Mountains located in the northwest and paralleling the coast. Next is the coastal zone, occupying the area between the mountains and the sea and including the delta of the Orinoco. And last is the *llanos* or plains area which includes the area between the mountains and the Orinoco River.

My evaluation of field sites in Venezuela was conducted during June 1989. At this time I visited one facility in the Coastal Range of north central Venezuela (Rancho Grande Biological Station), one facility near the area of the *tepuis* in the Guyana Highlands (Camp Canaima), and one facility located among the granitic outcroppings near Puerto Ayacucho in the Federal Territory of Amazonas (Camturama). Brief information about other lodges and facilities near these areas is also provided.

Rancho Grande

Contact

>Dr. Alberto Fernandez Badillo

Address

>Universidad Central de Venezuela (UCV)
>Facultad de Agronomia
>Apartado 4579
>Maracay 2101-A
>Venezuela

Description

Rancho Grande is a biological field station that is also open to birdwatchers, tourists, and other visitors as well as researchers. It has a very interesting history. Original construction on the building began in 1933, when it was planned to be a luxury hotel. However, this idea and construction ended in 1935 with the death of the country's ruler. The partially completed building was then abandoned and left to the forest.

During the 1940s, Rancho Grande was selected by the New York Zoological Society as the site for their headquarters for tropical research (William Beebe offers an excellent description of Rancho Grande and its wildlife in the book *HIGH JUNGLE* [see page 235]). In 1949, with the completion of the New York Zoological Society work, Rancho Grande was made a research station for the Ministry of Agriculture, and then later for the Ministry of Environment. It was then again abandoned.

Since 1966, the Rancho Grande Research Station has operated from a small portion of the actual building. At present, responsibility for the building is shared by the Universidad Central de Venezuela (UCV) and the Ministerio del Ambiente y de los Recursos Naturales Renovables (MARN). Rancho Grande is a name used to describe both the biological station and the national park in which it is located. Although the national park has been officially renamed to Henri Pittier (in honor of the botanist who conducted extensive research there), many people still use the name Rancho Grande to describe the building and the land.

VENEZUELA

Location

Rancho Grande is located slightly north of the city of Maracay in the north central state of Aragua. The station is situated in Venezuela's Coastal Range, a branch of the Andes that parallels the Caribbean coast. The building itself is located on the southern slope of a mountain at an elevation of approximately 1,100 meters and falls between 10°-11° North latitude.

Logistics

Most people arriving in Venezuela will do so by flying into Maiquetia Airport, located approximately 28km northwest of the capital city of Caracas. Immigration and customs checks were quick and without event during my visit. Taxis are readily available at the airport. One may either purchase a ticket for the taxi at a booth in the airport, or pay the driver directly. There are established fares depending on the time of travel (night or day), the section of Caracas one goes to, and the number and size of the pieces of luggage. I paid 400 B traveling at night to the Hotel Crillon (Gran Sabana area of Caracas), with a large duffle bag, a knapsack, and a briefcase. The drive from the airport to Caracas takes at least 20-30 minutes in light traffic.

From Caracas, it will be necessary to visit the UCV in Maracay, in order to pick up keys to the station, pay the fees, etc. Several alternatives are available to reach UCV Maracay, which is approximately 75 km southwest of Caracas. It is best to make contact with someone on the UCV faculty in order to determine which route is best and safest to travel.

Forest Type

The top of the Cordillera where the station is located has essentially undisturbed Cloud Forest. On the south side of the Cordillera this grades into rather seasonal and mostly somewhat disturbed Premontane Tropical Forest. On the north slope, there is a spectacular gradient through a Deciduous Tropical Forest down to a very dry, scrubby forest along the coast.

Seasonality

The dry season at Rancho Grande occurs approximately from the beginning of December to mid-April, reaching its driest point in February. The wet season begins in late April to early May and extends throughout November. The peak rainy months are July and August. The annual rainfall at Rancho Grande is approximately 1,660 mm.

Facilities

The Rancho Grande Biological Station consists of a large three floor cement building that was originally designed to be a luxury hotel. At least one half of the entire building was never completed. The field station itself consists of only approximately one fourth of the third floor. Efforts are currently underway to obtain the use of the other finished portions for research and educational purposes. The unfinished and unused sections of the building have suffered due to neglect and continuous encroachment by the forest, while the field station area remains completely serviceable.

Rancho Grande Biological Station in Henri Pittier National Park. (Near Maracay, Venezuela)

Access to the third floor of the building is gained through a wrought-iron gate adjacent to the garage area at the end of the driveway. One must climb two series of metal steps that have suffered greatly due to corrosion. Immediately to the right at the top of the steps is a set of double metal and glass doors that lead to the interior. The inner rooms consist of a kitchen, several lab rooms, a series of bedrooms or dormitories, a tool room, and a storage room.

The Rancho Grande building has the shape of a sickle or a question mark whose curved portion does not extend past its straight portion. The field station rooms are in the C-shaped portion of the building. The rooms are arranged off both sides of a wide hallway about 35 meters by 4 meters. It contains a gas-powered stove and refrigerator and has a sink with potable water supplied by a nearby mountain stream. All visitors must currently bring their own food to Rancho Grande. The pots, pans, glasses, silverware, plates, etc. are available at the station. There are tables to seat 10 in the kitchen, but a wide counter that runs almost the entire length of the kitchen perimeter could easily accommodate another 15 people.

There are two small and one large dormitory at Rancho Grande. The smaller bedrooms accommodate three single beds each. Each smaller bedroom has a large attached bathroom with a flush toilet, sink, and shower. The same cold mountain stream water is used in the sink and shower. The dormitories may contain small night stands, but are lacking in other amenities. Like all the other rooms, the floors are of poured concrete. The beds are simple metal folding structures equipped with thin foam mattresses. Blankets and pillows are also provided. However, visitors must provide their own soap and towels. The large dormitory holds approximately 25 beds and has its own bathroom.

There are two laboratory rooms of approximately the same size as the kitchen (5 x 7 meters). They are also each equipped with a sink and with counter space extending around the entire perimeter. Researchers should bring whatever scientific equipment they need with them.

All rooms have electrical outlets, as well as the outer areas. Electricity during my stay was provided by a portable gasoline-powered generator. The tank and power were sufficient to run several room lights simultaneously for 6-7 hours. A much larger,

Rancho Grande Biological Station in Henri Pittier National Park. (Near Maracay, Venezuela)

permanent generator exists near the garage area, but is in need of repairs. I was told these repairs would be made in the near future. The outlets accept the typical two parallel flat-pronged plugs used in the United States.

There is a wide (approximately 3 meters) covered balcony that runs adjacent to the field station rooms. This provides an excellent vantage point for birdwatchers to scan the surrounding trees, or for entomologists to place light traps at night. There is also three times as much additional balcony or roof area that is uncovered. On this flat, uncovered portion is the metal framework (6 x 4 meters) of what appears to have been a greenhouse. This could easily be converted again to a greenhouse, a screenhouse or screened-in laboratory.

Trail System

There is a short 'loop' trail that extends for about 1-2km behind the Rancho Grande building. Two other trails branch off from the loop trail near its beginning. One goes to Pico Rancho Grande and the other goes to a dam. There is a map available of the loop trail, although there are no markers along it.

A short walk from the field station is the famed 'Paso Portachuelo' or Portachuelo Pass. This is a narrow cut in the mountain ridge and is the site of heavy migrations of both birds and insects. To reach the Portachuelo Pass, one walks down the driveway to the Maracay-Ocumare Road and turns right to follow the road in the direction of Ocumare. A metal gate on the left side of the road, located shortly after the Rotary Club sign, is reached about 5 minutes from the station. About 20 meters through this gate and along a heavily weeded path, is Portachuelo Pass. If one continues over the pass (about 8-10 meters) there is a trail on the other side that goes up the mountain ridge to Pico Periquito.

Costs

The cost to stay at Rancho Grande and use the facilities (kitchen, generator, etc.) was $10 per night during my stay (June 1989). This minimal amount may soon be increased (probably to $15 per night) in an effort to generate income to use for improving facilities, and repairing and maintaining those that already exist. One must also plan on the indirect costs of transportation to reach Maracay or Rancho Grande, and of purchasing food.

Comments

The Rancho Grande Biological Station is a facility with vast potential. It is easily accessible by a well-paved road from Maracay, and is located in one of the most beautiful cloud forests in the Neotropics. Currently, only about 10% of the actual building space is in use. However, should the Universidad Central de Venezuela be given full control of the building, that may quickly change.

The accommodations at Rancho Grande are not luxurious, but are more than adequate for field biologists or serious naturalists and birdwatchers. Stairs are rickety, the roof leaks in places, and many windows are broken, but Rancho Grande still retains some of its charm from earlier times. The rather dismal impression created by the bleak appearance of the unused part of the building gives way to a feeling of awe to know that here is where some of the most famous tropical biologists in the world worked 40 years ago.

The forest around Rancho Grande appears to be primary growth and little disturbed. Many large buttress trees are present, as are great quantities of bromeliads and other epiphytes. The trail system is mapped and easy to follow, but at present only consists of one short loop and two straight trails. Howler monkeys are present and their vocalizations can be heard daily. Over 500 species of birds have been identified from Rancho Grande, which represents 40% of the species found in Venezuela and 7% of the species worldwide. Many species of birds are easy to view from the balcony or roof area in the trees growing next to the building.

Rancho Grande is located only 12 km from the Universidad Central de Venezuela, which makes it an ideal selection for researchers planning long-term projects. The UCV is a university of high quality and is extremely well equipped. Some of the most eminent and knowledgeable biologists in Venezuela belong to its faculty. The UCV also possesses some of the finest flora and fauna collections in the country. Staff at the UCV were most helpful during my visit and are actively trying to encourage the use of Rancho Grande by all interested parties (scientists, tour groups, birdwatchers, naturalists, etc.). Although not entirely unknown to researchers and birdwatchers, Rancho Grande is used only infrequently by international visitors at present. If sufficient funding is obtained to repair, maintain, and improve the facilities there, it may well become one of the premier sites for tropical biology and international tourists and naturalists.

Canaima

Canaima is a national park located in the southeastern part of Venezuela in the State of Bolivar. It is in an area that forms part of the Guyana Highlands and is famous for its towering flat-topped mountains (called *tepuis*) and Angel Falls (the world's highest waterfall). There are two tourist facilities at this national park. One is Camp Canaima, which is run by AVENSA Airlines. The other is Camp Ucaima, which is run by Jungle Rudy. During my evaluation of the area I stayed at Camp Canaima, but took one afternoon to visit Camp Ucaima as well.

VENEZUELA 119

Camp Canaima

Contact

Hoturvensa

Address

800 Brickle Avenue
Suite II 1109
Miami, FL 33131 USA
Telephone: (800)-872-3533 (AVENSA Airlines)

Description

Camp Canaima is a tourist facility which provides lodging and food to those visitors wishing to visit Canaima National Park in the Guyana Highlands of east central Venezuela.

Location

Camp Canaima is located only a half kilometer from the Canaima landing strip in the State of Bolivar. It is situated at approximately 400 meters elevation between $6°$-$7°$ North latitude.

Logistics

Most (if not all) of the people visiting Canaima are booked into some type of 'package' which includes airfare to Canaima, and food and lodging at the Camp. By booking my stay from the U.S. I received from my travel agent a 'miscellaneous charges order' (M.C.O.). This voucher was then given to an AVENSA agent at Maiquetia Airport outside of Caracas, before my flight to Canaima. The AVENSA representative then returned to me a different voucher which was to be used to obtain my airline ticket, and food and lodging at the Camp. This transaction was conducted at the portion of the AVENSA ticket counter in the National terminal marked 'Venta de Boletos - Contado y Credito'.

AVENSA is the only commercial passenger airline that flies to Canaima. The usual route is from Caracas to Ciudad Bolivar (approximately 45 minutes flying time), then Ciuidad Bolivar to Canaima (approximately 25 minutes flying time). Multilingual representatives of the camp meet visitors at the landing strip, herd them into awaiting open-air trucks with benches to sit on, and quickly explain the registration process. Aside from taking certain biographical information, they also take your luggage claim checks (numbered adhesive labels that are stuck onto your ticket). This allows them to collect your baggage.

The truck ride to Camp Canaima takes only 2-3 minutes. You disembark near the registration desk and then wait in line to hand to the management the form your camp representative has already filled out on the truck. Upon receiving the form, the Camp management will give you a room key, a meal voucher (stamped with exactly the number of meals you will need while at the Camp), and a voucher entitling you to a boat ride around the adjacent lake. Your 'guide' will then usually take your baggage to your room and ask you what excursions you want to sign up for.

Forest Type

One does not have immediate access to true rainforest at Camp Canaima. Some of the hillsides are visibly forested, but the only easily accessible trail through forest habitat is at the end of the Canaima road, between Camp Canaima and Jungle Rudy's Camp Ucaima. This trail only lasts for approximately 0.5 km. Canaima National Park does contain tropical forest, as is visible between the *tepuis* on the flight to Canaima. However, Camp Canaima is surrounded by a vegetation classed as 'chaparal'.

Seasonality

The dry season normally occurs from December or January to March and April. The rainy season takes place from May to November. the months of April and November are usually transition months.

Facilities

Camp Canaima offers over 100 rooms in a series of stucco-walled cottages, cabins, and bungalows. Ceilings were of wood and roofs were covered with some type of plant fiber. The number of beds in different rooms varied. I stayed in one side of a two-room cottage located on a shallow hillside amongst mango trees. The main room contained three beds, had a tile floor, and screened jalousy windows. There were two electric lights and outlets accepting a two parallel-pronged plug. There was a small table with a thermos of drinking water, shelves and hangers for clothing, and shelves behind the beds. The bathroom contained a cold water shower, sink, and flush toilet. The sink and shower water were not potable. Towels and soap were provided.

The dining room was a large open-air structure with a high conical roof of plant fiber. The seating capacity was approximately 125-150. Meals were served cafeteria style, with clients taking trays and moving past a counter where they selected verbally from the limited menu of the day. Meals were included in the tour packages, but all beverages other than tea and coffee were separate and had to be paid for when they were brought. The food in general was good. Meals were served at the following times: breakfast (7:00-9:00 am), lunch (12:00-2:00 pm), and dinner (7:00-9:00 pm).

Hillside guest cabin at Camp Canaima. (Canaima, Venezuela)

Tepuis and waterfalls visible from the dining hall of Camp Canaima. (Canaima, Venezuela)

Most of the facilities at Camp Canaima were centrally located, with the exception of the cabins which were scattered from the beach along the lake to the surrounding hillside. The dining hall, registration desk, and AVENSA office were located on one side of the main (only) road, while an open-air covered structure, housing tourist agencies, was immediately opposite on the other side. The tour agency facility actually consisted of little more than four desks and some displays of photographs taken in the area. A variety of excursions were offered from Camp Canaima ranging from half day to several days in duration. Most excursions entailed both hiking and boat transportation, with some including jeep or truck travel as well. There were also excursions involving flights in light aircraft, which were extremely expensive.

Trail System

There was a series of trails through the 'chaparal' on the far side of the lake that bordered the beach and dining hall. However, trails were not marked and maps were not available. All excursions at Camp Canaima were done through tour guides. The nearest one could get to hiking on one's own was to follow the Canaima road up past the kitchen and towards the waterfalls.

After a half kilometer or so it will lead to a path through the forest that ascends up to the location of Camp Ucaima. It took approximately 45 minutes to walk the road and the path to reach Camp Ucaima from Camp Canaima.

Costs

My tour package at Camp Canaima was booked in the United States, directly through an AVENSA agent by phone. I paid $153 for three days and two nights. This fee included an additional $18 charge for using a triple room. After booking my reservations with AVENSA, I then had my regular travel agent call them so that the travel agency could provide me with the M.C.O. (miscellaneous charges order) voucher needed. By calling AVENSA directly I saved approximately $40 from the price for the same package that I was quoted by a travel agent.

For most traveling of this nature in Venezuela, it is even cheaper to make your bookings after arrival. Of course, this may be an inconvenience or an intolerable uncertainty for people with tight schedules. However, for those people traveling on a tight budget and flexible schedule, it may provide a means for saving some money. Unfortunately, the economic situation in Venezuela has led to price increases on almost everything.

During my stay at Camp Canaima, examples of half day tours cost 600 B ($15) and 1,000 B ($25), while some full day and longer excursions cost 2,000 B ($50) and 8,000 B ($200). A 45 minute tourist flight in a light plane cost 1,800 B ($45) per person, based on a five person group.

Two of the tour agencies operating from Camp Canaima are:

Canaima Tours
Hortuvensa
Parque Nacional Canaima
Canaima, Estado Bolívar
Venezuela

Excursiones Churum Vena
Parque Nacional Canaima
Canaima, Estado Bolívar
Venezuela

Comments

Camp Canaima is a tourist-oriented facility that accommodates a large number of visitors on a daily basis. Some tourists come for only one night, although two or three is more typical. Unfortunately, with the exception of Jungle Rudy's Camp Ucaima, Camp Canaima is the only facility available to visitors wishing a quick look at Angel Falls and the *tepuis*. I felt 'hustled' from the moment I got off the plane until I left. Service could have been much better in all respects, and all employees except the tour guides seemed to have a lackadaisical attitude. While the rooms were comfortable, the housing was shabby and in need of several minor repairs. The open-air dining hall afforded a spectacular view, but made dining a constant battle between visitors and flies.

In support of Camp Canaima, it gives visitors easy and rapid access to one of the most interesting and unique areas of the world. It permits tours and excursions in relative comfort to what otherwise would almost be totally inaccessible sites. Even the flight into Canaima from Caracas normally provides an excellent view of the *tepuis* and Angel Falls. The location of the camp permits swimming and sunning on a small beach within a view of spectacular waterfalls overshadowed by *tepuis* in the background.

Camp Ucaima

Jungle Rudy's Camp Ucaima is the only local competition for AVENSA's Camp Canaima at the Canaima National Park. Camp Ucaima is located approximately a half hour by road from the Canaima airstrip. I visited Camp Ucaima only briefly one afternoon, but received additional information from Jungle Rudy's brochure.

Jungle Rudy is a Dutch ex-patriate who has spent approximately 30 years in Venezuela. The Ucaima Camp which he established is much smaller than Camp Canaima and can accommodate about 20-30 people. My cursory examination of his facilities revealed a very well maintained and 'homey' type of establishment. There was a small lodge building with an open-air bar, lounge area, and dining hall. The walls of the lounge were decorated with local Indian artifacts. The guest room I viewed was very neat and modern in appearance.

Jungle Rudy offers his own excursions to tourists based on his many years of guiding experience. I assume they are similar in nature to those offered by the other tour companies.

Camp Ucaima appeared to me to be a quiet, intimate, and tranquil type of place, as opposed to the bustling activity of at least dozens of tourists that come and go everyday at Canaima. The two facilities present very distinct alternatives for visiting the same general area. For additional information on Camp Ucaima, contact Jungle Rudy at the following address:

Camp Ucaima - Jungle Rudy
Apartado 61.879
Caracas 1060-A
Venezuela
Telephone: 661-91-53
Telex: 26103 NOTIC

Camturama Amazonas Resort

Contact

Olivia Blanco

Address

There was no postal address available for Camturama at the time of my visit. However, Olivia Blanco could be reached for information and reservations at the following Caracas phone number: 979-58-21.

Description

Camturama is a tourist lodge and resort that opened for business late in 1988.

Location

Camturama is located approximately 20 km south of the city of Puerto Ayacucho in the Federal Territory of Amazonas. It is situated between 5°-6° North latitude on the bank of the Orinoco

River in southwest Venezuela along the Colombian border. It is at approximately 200 meters elevation.

Logistics

Most visitors to Camturama will fly into the airport at the town of Puerto Ayacucho. AVENSA maintains daily flights to and from Caracas in Boeing 727 jets. Representatives of Camturama will meet visitors at the airport and transport them to the resort in either their very modern bus or one of their jeep vehicles. The drive from the airport takes approximately 30 minutes.

Forest Type

Camturama is located in an area of savannah that is dotted with large and small black granite hills. Some of these hills (*cerros*) are covered with forest. Also accessible from the resort is gallery forest along the Orinoco and other rivers.

Seasonality

The rainy season in the vicinity of Camturama usually begins in May and extends to the end of October. The dry season runs from November to about the end of April.

Facilities

The buildings at Camturama were all extremely modern, having been constructed within the past two years. Construction was of brick with glass windows. Roofs were of wood, but covered with thatching to give the appearance of native architecture. My room, which appeared identical to the majority of the others, had the following characteristics.

The cabin unit was square and housed four guest rooms; two to a side with doors opening opposite each other in the middle. The flooring was of tile, with stucco walls and glass windows covered by heavy curtains. Only one small window section was jalousied and screened while the others were immovable. The room was efficiently air conditioned by a large wall unit. There were two double beds, comfortable and homey in appearance. In one corner was a large, open built-in closet with hangers and shelves. The walls were decorated with traditional *artesania* of the area. There were two end tables, a larger vanity sized table, a chair, and a mirror. The private adjoining bathroom was large and completely tiled. There was a shower with 24 hour hot water, a sink with potable water from a well, and a flush toilet.

There was both an overhead electric light and an electric lamp. There were two electrical outlets that accepted both a plug consisting of two flat, parallel, metal prongs of equal size, and two parallel, cylindrical prongs. Electricity was 110V.

Approximately 32 rooms of this type currently existed at Camturama. Under construction were 28 single bedrooms. These were to be housed in brick and glass buildings of round shape with conical roofs. This would bring the total capacity of the resort to about 90 persons.

The dining hall was an equally impressive glass and brick building. It, too, was equipped with electricity which powered overhead lights, four overhead fans, and a huge air conditioning unit that was used during meal times. The seating capacity was approximately 75. Serving was done buffet style with a staff of kitchen helpers dishing out the food. All meals were excellent in quality and quantity. The menu was varied and consisted of both traditional Latin American dishes and European and American standards. A variety of fresh tropical fruit juices and soft drinks were available, as well as beer and liquor from the bar.

Beyond the dining hall was a combination bar and discotheque, built on and around one of the black granite rock outcroppings so common in the area. Adjoining the bar was a separate game room with pool tables and other recreational games. Both bar and game room adhered to the brick and glass construction, were comfortably furnished and attractively decorated, and were air conditioned.

Planned for the future were a reception area and administrative office, two shops, a swimming pool, and two tennis courts. There were several open-air, covered patios for sitting and talking, a set of restrooms located between the dining hall and game room, and an artificial lagoon with a covered, open-air patio. All of the buildings were interconnected by a fieldstone sidewalk and the camp was bisected by a red gravel driveway crossed by a network of grates and an underground drainage system.

Trail System

There was no trail system established in the areas of forest I examined closest to the resort. A guided excursion is possible to the nearest forest covered hill (about a 10 minute walk), where a single short (<1km) overgrown, unmarked and unmapped trail exists.

Costs

A 4 day/3 night package booked through an agency in the United States cost me $302. This included round trip airfare from Caracas to Puerto Ayacucho, transfers between Camturama and the Puerto Ayacucho airport, food and lodging at Camturama, and guided excursions during my stay. According to the

Modern guest cabins at the Camturama Amazonas Resort. (Near Puerto Ayacucho, Venezuela)

Camturama manager, approximately 20% can be saved by booking reservations directly through the Caracas number (provided earlier). I was also informed that visitors not interested in excursions would be able to pay a basic daily room rate (800 B = $20) and a standard fee for each meal: breakfast (150 B = $3.75), lunch (250 B = $6.25), and dinner (250 B = $6.25). These prices were all quoted during June 1989. All plans and prices should be confirmed through the Camturama Amazonas Resort management well before arrival.

Reservations to stay at Camturama can be currently made through the following U.S. travel agent:

Lost World Adventures
1189 Autumn Ridge Drive
Marietta, GA 30066 USA
Telephone: (404)-971-8586

Comments

Camturama provides extremely comfortable accommodations and offers a variety of interesting excursions. Among these are visits to the petroglyphs at Cerro Pintado, a trip along the Orinoco with various stops, a visit to the ethnological museum and Indian market in Puerto Ayacucho, and a trip through nearby forest. As stated earlier, Camturama is located in savannah and the forests available are the gallery forests along the Orinoco and other rivers, and the forested slopes of the various sized granite rock hills that abound in the area. There is no immediate access however to a large area of continuous tropical rainforest.

Manaka Lodge

The Manaka Lodge is a fishing camp and tourist lodge located in the Amazonas Federal Territory. For additional information contact:

Otto Winckelmann
Apartado 76463
Caracas, Venezuela
Telephone: 959-15-70

Hato Piñero

Hato Piñero is a tourist lodge and ranch located in the *llanos* or plains region of Venezuela. For more information, write to:

Bio Tours
GBS C.A.
Attn: Ani Villanueva
Poba International #156
P.O. Box 52-1308
Miami, FL 33152-1308 USA

Caracas

Caracas is a modern city of more than four million inhabitants, located on the north central coast of Venezuela. It is sprawled among the mountains of the Coastal Range at an elevation ranging from 850-950 meters. Although depressed oil prices and other factors have caused great economic problems recently, the signs of obvious wealth and affluence still abound in this city.

During my time in Caracas I stayed at the Hotel Crillon, located on the corner of Avenida Las Acacias and Avenida Libertador in the Sabana Grande area or district. The Hotel Crillon is a large, modern hotel with 13 stories and 80 rooms. Each room contained two beds, a private bathroom with modern sink, toilet, bidet, and 24 hour hot water shower, and a small, separate foyer with couch, small refrigerator, and television. All rooms were air-conditioned and contained ample closet and drawer space, as well as a small writing area. Each room also had a telephone.

The accommodations were quite comfortable, with my only complaint being the phone system. Often the phones in the room did not function correctly, and the several times when people tried to call the hotel from the U.S., they could not get through. This may not be through any fault of the hotel. I placed several calls with little or no difficulty from the hotel to the United States. One should be aware however, that phone calls placed to U.S. through hotels are very expensive. My calls calculated out to costing between 150 B-200 B (approximately $3.75 to $5.00) per minute! There was no charge for calls made within Caracas.

The Hotel Crillon management working at the front desk all spoke English, and were extremely courteous and helpful. My cost as a single for the room described above was 950 B plus 95 B tax for a total of 1,045 B (approximately $26). Other cheaper hotels may well be available in the same general area. Within two blocks of the Hotel Crillon was the main boulevard and shopping district called the Real De Sabana Grande. This street, which is closed to traffic through much of its course, has many fine shops, bookstores, restaurants, open-air cafes, and fast food places. Many vendors also set up small tables and market items directly on the street. Also located two blocks from the Hotel Crillon is the Plaza Venezuela station of the Metro underground train.

The address of the Hotel Crillon is:

Hotel Crillon
Avenida Libertador con Avenida Las Acacias
Caracas 105, Venezuela
Telephone: 71-69-12   71-44-11
             71-69-13   Or   71-44-12
             71-69-14   71-44-13
                                71-44-14

Telex: 21203
Cable: HOCRI

Like most hotels in Caracas, the Crillon will only hold a reservation until 6:00 pm, unless it has been paid for in advance. The hotel has its own taxis, which can be arranged for prior to their need. The cost of a taxi from the Hotel Crillon to the Maiquetia Airport was 300 B (approximately $7.50) during the day and 400 B (approximately $10) at night.

Traffic in Caracas is usually extremely heavy during the day. Although the airport can be reached in 20-30 minutes in light traffic, I always allowed at least an hour. For most travel within the city, the subway or Metro serves very efficiently. It operates in a similar manner to the Metro of Washington, D.C. in the United States. One purchases a ticket from a machine by pushing a button to select the ticket value and then feeding in the corresponding amount of coins. The ticket will then be dispensed to a small chamber where you pick it up. This ticket is then placed in a slot of a turnstile to enter the actual train area. The ticket passes through the turnstile and pops up the top, where the passenger must collect it. This same ticket will be used in the same way by travelers when they exit the station of their destination. If one does not have the correct amount of coins needed, tickets can be bought at a booth. The lines for this, however, are usually long. The amount needed for a ticket can be determined from a map of the Metro stations located at each ticket vending machine, and on the walls of the Metro station. Of course, to use it efficiently, one must know which Metro station is closest to their destination.

Comments On Venezuela

Venezuela offers both tourist and biologist an excellent variety of ecologically distinct sites with facilities. These include areas with rainforests, plains (*llanos*), *tepuis*, and beaches. Tourism only appears well developed with respect to Canaima and Angel Falls, and some areas in the *llanos*. Venezuela has many fine scientific institutions and organizations that have conducted research and exploration of the indigenous flora and fauna. Several active conservation organizations also exist and are responsible for supporting research and increasing public awareness of endangered flora and fauna.

Books

There are a number of bookstores or *librerías* located along the boulevard Real De Sabana Grande. Most will offer the beautifully illustrated 'coffee table' type books on Venezuela. Some will also have books on various flora and fauna (exs. fruits, orchids, snakes, diurnal butterflies). Excellent volumes on specific flora and fauna have been published by the Sociedad De Ciencias Naturales La Salle (La Salle Society of Natural Sciences).

The best bookstore in Caracas for books on birds, flora and 'na, natural history, biology and ecology is that operated by the 'dad Conservacionista Audubon de Venezolana (Venezuelan

VENEZUELA 133

Audubon Society). Their office is located in the Paseo Las Mercedes (a large shopping center or mall) in the Las Mercedes area of Caracas. This mall or commercial center is in the same building as the Holiday Inn and is located across from the Tamanaco Hotel. The office of the Venezuelan Audubon Society is small and out of the way, located at the end of the mall furthest from the front entrance, past an ice cream store and along side of an exit to an underground garage.

Another specialized bookstore with nature books is the Libreria Ecologica, which is located at the following address:

Fundación de Educación Ambiental
Libreria Ecológica
Centro Simon Bolivar
Edificio Camejo
Planta Baja
Local 7
Esquina de Camejo
Caracas 1010

MARIPOSAS DIURNAS DE VENEZUELA - INTRODUCCIÓN A SU CONOCIMIENTO
Alvarez Sierra, José Ramón and José Ramón Alvarez Corral
1984.

HIGH JUNGLE
Beebe, William
1949. Duell, Sloan and Pearce

RORAIMA - THE CRYSTAL MOUNTAIN
Brewer Carias, Charles
1988. Editorial Arte (Caracas)

VENEZUELA
Brewer-Carias, Charles
1988. Editorial Arte (Caracas)

ALGUNAS PLANTAS USADAS EN LA MEDICINA EMPÍRICA VENEZOLANA
Chitty, Francisco Delascio
1985. Litopar C.A. (Caracas)

THE LOST WORLD
Doyle, Arthur C.
1912. Hodder and Stoughton

LAS ORQUÍDEAS DE VENEZUELA
Dunsterville, Galfrid C.K.
1987. Craficas Armitano, C.A. (Caracas)

VERTEBRATE ECOLOGY IN THE NORTHERN NEOTROPICS
Eisenberg, John F. (Editor)
1979. Smithsonian Institution Press

IMPRESSIONS OF VENEZUELA - IMPRESSIONES DE VENEZUELA
Fischer, Wenzel and Willy Haas
1977. Publicaciones Fher, S.A. (Bilbao, Spain)

GUÍA DE LOS PARQUES NACIONALES Y MONUMENTOS NATURALES DE VENEZUELA
Fundación De Educacion Ambiental
1982. Ediciones Fundación De Educación Ambiental

BIRDING IN VENEZUELA
Goodwin, Mary Lou
1987. Sociedad Conservacionista Audubon De Venezuela

VENEZUELA ALIVE
Greenberg, Arnold
1989. Alive Publications

PLANTAS ORNAMENTALES DE VENEZUELA
Hoyos F., Jesús
1982. Sociedad De Ciencias Naturales La Salle. Monografía No. 31 (Caracas)

GUÍA DE ÁRBOLES DE VENEZUELA
Hoyos F., Jesús
1987. Sociedad De Ciencias Naturales La Salle. Monografía No. 32 (Caracas)

FRUTALES EN VENEZUELA
Hoyos F., Jesús
1989. Sociedad De Ciencias Naturales La Salle. Monografía No. 36 (Caracas)

LA SELVA NUBLADA DE RANCHO GRANDE - PARQUE NACIONAL "HENRI PITTIER" - EL AMBIENTE FISICO, ECOLOGÍA VEGETAL Y ANATOMÍA VEGETAL
Huber, Otto (Editor)
1986. Fondo Editorial - Acta Cientifica Venezolana and Sequros Anauco C.A. (Caracas)

LOS PARQUES NACIONALES DE VENEZUELA
Instituto Nacional De Parques (INPARQUES)
1983. Fundación De Educación Ambiental (Caracas)

DEL RORAIMA AL ORINOCO - TOMO I
Koch-Grunberg, Theodor
1981. (Originally published in German in 1917) Ernesto Armitano, Editor
(Caracas)

SERPIENTES DE VENEZUELA
Lancini V., Abdem R.
1979. Ernesto Armitano, Editor (Caracas)

A GUIDE TO THE BIRDS OF VENEZUELA
Meyer de Schauensee, R. and W.H. Phelps, Jr.
1978. Princeton University Press

MARAHUAKA
Michelangeli Ayala, Armando and Fabian Michelangeli Ayala, Reinaldo Stevie Borges, Walter Smitter, Armando Subero, and Klaus Jaffe.
(Available through Terramar Foundation. See page 140.)

IN TROUBLE AGAIN: A JOURNEY BETWEEN THE ORINOCO AND THE AMAZON
O'Hanlon, Redmond
1989. The Atlantic Monthly Press

GARDEN PLANTS OF THE TROPICS (VENEZUELA)
Oliva-Esteva, Francisco
1986. Ernesto Armitano, Editor (Caracas)

VENEZUELA Y SU GEOGRAFÍA - REGION DE GUYANA
Ovelar, Silvio R. (Editor)
1989. Actividades Educativas EDUDACTA C.A.

VENEZUELA LEJANA
Vareschi, Volkmar
1975. Ernesto Armitano, Editor (Caracas)

VENEZUELA - ESPAÑOL, ENGLISH, DEUTSCH
Weidmann, Karl
1987. Oscar Todtmann Editores

PÁRAMOS VENEZOLANOS
Weidmann, Karl
1988. Oscar Todtmann Editores

Maps

Road maps of Venezuela, and combination maps and tourist guides can usually be purchased in most bookstores. One map that covered the roads of northern and central Venezuela, with an excellent plan of Caracas on the reverse side is the *Mapa De Carreteras De Venezuela Plano De Caracas Y Litoral Central*. The best road map I found covering the entire country was the: *Guia Progreso - Mapas De Carreteras Para El Turismo En Venezuela*, published by Seguros Progreso S.A., 1987. This large, spiral bound booklet was available in a bookstore called Las Novedades, located in a small mall on the Real De Sabana Grande near the Chacaito station of the Metro.

Detailed physical-political maps of Venezuela were difficult to obtain while in Venezuela. Maps covering some areas of the country were available at the Libreria Ecologica (see page 133). Others, including a vegetation map and beautiful wall-sized relief map were available from the Cartographic Office, located above the Libreria Ecologica. The location for this office is:

Centro Simón Bolivar
Avenida Este 6 Esquina Camejo
1er Piso (First Floor)

Another wall-sized map showing both terrain and cultural landmarks was available in some of the shops in the airport.

Domestic sources of maps of Latin American countries are listed on pages 290 and 291.

Tourist Information Sources

Departamento De Turismo
c/o Central Office of Information
Palacio de Miraflores
Caracas
Venezuela

Corporación De Turismo De Venezuela (Corpoturismo)
Centro Capriles 7°
Plaza Venezuela
Apartado 50200
Caracas
Venezuela
Telephone: 7818370   Telex: 2396

Corporación Nacional De Hoteles Y Turismo (CONAHOTU)
Apartado 6651
Caracas
Venezuela

Tourist/Travel Agencies

Lost World Adventures
1189 Autumn Ridge Drive
Atlanta, GA 30066 USA
Telephone: (404)-971-8586   Telex: 626-27-070

Jorge M. Gonzales
Edificio Don Luis, Apartado 501, Piso 5
Esquina Horcones, El Conde
Caracas D.F. 1010
Venezuela

Carlos Rivero Blanco Nature Tours
Apartado 63011
Chacaito
Caracas 1067-A
Venezuela
Telex: 278999 ECUVE

Tobogan Tours C.A.
Pepe Jaimes A.
Avenida Rio Negro No. 44
Puerto Ayacucho
T.F. Amazonas
Venezuela
Telephone: 21700

VENEZUELA 139

Turven Tropical Travel Services
Calle Real De Sabana Grande
Edificio Unión
Local 13
Apartado 60627
Caracas, Venezuela
Telephone: 717098 or 728202
Telex: 21213 TURSA

Conservation Organizations

Sociedad Conservacionista Audubon De Venezuela
Apartado 80450
108 Caracas
Venezuela
Telephone: 913813

BIOMA
Fundación Venezolana Para la Conservación De La Diversidad
  Biológica
Apartado 1968
1010-A Caracas
Venezuela
Telephone: 5718831 or 5716009 or 5719113

FUDENA
Fundación Para La Defensa De La Naturaleza
Apartado 70376
Caracas 1011-A
Venezuela
Telephone: 2393272 or 359454
Telex: 23280 HIDECVC

Note: FUDENA has published an excellent work titled: *VENEZUELA - TOWARDS A NATIONAL CONSERVATION STRATEGY: SPECIES CONSERVATION ACTION PLAN - 1988-1992.* This publication explains in both English and Spanish the current conservation goals and projects of FUDENA.

Sociedad Conservacionista Mérida
Apartado 552
Mérida
Venezuela
Telephone: 440576 or 449461

Scientific Organizations/Institutions

Sociedad De Ciencias Naturales la Salle
Apartado 1930
1010-A Caracas
Venezuela
Telephone: 7824845 or 7828711
Telex: 21553 FLASA - VE

Sociedad Venezolana De Ciencias Naturales (SVCN)
Apartado 1521
1010-A Caracas
Venezuela
Telephone: 217653 or 217780

Fundación Terramar
Universidad Simón Bolivar
Apartado 89000
Caracas 1080
Venezuela
Telephone: 9621201
Cable: UNIBOLIVAR

Instituto Venezolano De Investigaciones Cientificas (IVIC)
Apartado 1827
Caracas 1010A
Venezuela
Telephone: 6811188 or 691951
Telex: 21338

Universidad Central De Venezuela
Facultad De Agronomía
Instituto De Zoología Agricóla
Apartado 4579, Codigo Postal 2101-A
Maracay, Aragua
Venezuela

Sociedad Venezolana De Ecología
Apartado 47543
Caracas 1041-A
Venezuela

CHAPTER 1E TRINIDAD

General Information

Area: 4,828 sq km (1,864 sq miles)

Population: 1,261,000 (Trinidad and Tobago)

Capital: Port-of-Spain (Population: 57,000)

Language: English

Currency: 1 Trinidad and Tobago dollar (TT$) = 100 cents

    Coins: 1, 5, 10, 25, and 50 cents
    Notes: 1, 5, 10, 20, and 100 dollars
    Exchange Rate: U.S. $1 = 4.25 TT$ (June 1989)
                  U.S. $1 = 4.25 TT$ (March 1990)

National
Airlines: BWIA (British West Indies Airlines)

Telephone Trinidad 809
Code: No special city codes are needed.

Trinidad and Tobago

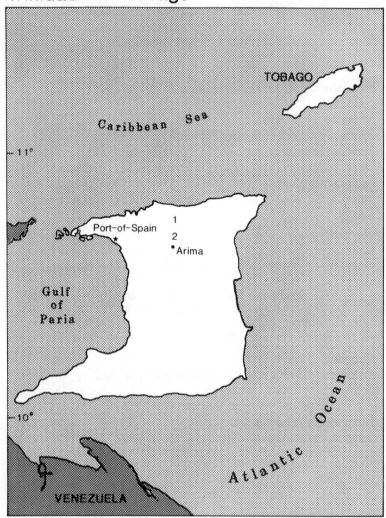

1 - Asa Wright Nature Centre and Lodge 2 - Simla Research Station

Introduction

Trinidad is the southernmost island of the Caribbean Sea, located approximately eight miles from the Venezuelan coast, opposite the mouth of the Orinoco River. It is politically joined and forms a republic with Tobago, a small island approximately 20 miles to its northeast. Trinidad is a small island about the size of the state of Delaware.

Trinidad is a rectangular shaped island with peninsular spurs of land jutting out from the southwest, northwest, and northeast corners. Three mountain ranges traverse the island, none of which have peaks greater than 1,000 meters. The Northern Range runs east and west, continuing the line of the Andes in Venezuela. The Central Range runs diagonally across the central part of the island from southwest to northeast. The Trinity Hills, from which the island was named, exist in the southeastern corner. Areas outside of the ranges are generally low, with many swamps along the coastline.

My evaluation of rainforest sites in Trinidad is based primarily on observations made and information obtained during June 1989. At this time I visited two facilities in the Arima Valley of Trinidad's Northern Range (Asa Wright Nature Centre And Lodge and the Simla Research Station). Brief information is also provided on Victoria Regia, a new research station that has opened in central Trinidad.

Asa Wright Nature Centre And Lodge

Contact

    Ian Lambie

Address

    Asa Wright Nature Centre
    P.O. Bag 10
    Port-of-Spain
    Trinidad, West Indies

    Port-of-Spain Office
    Telephone: (809)-667-4655

    U.S. Representative:
    Caligo Ventures
    387 Main Street
    P.O. Box 21
    Armonk, New York 10504-0021
    Telephone: (914)-273-6333 (Local)
                   (800)-426-7781 (Toll Free)

Description

    The Asa Wright Nature Centre and Lodge (AWNCL) is a protected conservation area with facilities for tourists, naturalists, and birdwatchers. Comfortable accommodations and meals are provided in an 80 year old estate house and adjacent cottages. In recent years, the AWNCL has been used primarily by birding groups and individual birders.

    During the 1940s the property was known as the Springhill Estate and contained plantations of cacao, coffee, and citrus. It was purchased by Newcome Wright and his wife Asa in 1947. The property received a lot of attention from scientists of the New York Zoological Society during 1950-60, who were headquartered at Simla (see pages 151-157) a short distance away. The Asa Wright Nature Centre was founded in 1967 to preserve the wildlife on the property (especially a colony of oilbirds) and to establish a recreation and study area. It has functioned in that capacity to date.

Location

The Asa Wright Nature Centre and Lodge is located approximately 12 km north of the town of Arima at an elevation of approximately 360 meters in the Arima Valley of the Northern Range. This is in the north central part of the island. Trinidad lies between 10° and 11° North latitude.

Logistics

The Asa Wright Nature Centre and Lodge is approximately 45 km by road from Port-of-Spain and about 25 km by road from Piarco Airport. When traveling from Port-of-Spain, one can take either the Eastern Main Road or the Churchill Roosevelt Highway east until reaching the town of Arima. The Eastern Main Road goes directly through Arima, while travelers on the Churchill Roosevelt highway may follow either O'Meara Road or Tumpuna Road north to reach the town. Once in Arima, one continues in a northward direction, traveling uphill until reaching Blanchisseuse Road. The AWNCL is located approximately 12 km north of Arima. The entrance to the centre is on the left (west) side of Blanchisseuse Road, shortly after the 7 1/2 mile marker (which is a white post with a black 7 with two lines under it).

When traveling from Piarco Airport, visitors may follow Golden Grove Road north to reach either Churchill Roosevelt Highway or the Eastern Main Road. Turning east onto either of these routes, the directions will be the same as above.

Note: Driving in Trinidad can be extremely hazardous. Vehicles use the left side of the road and 'maxi-taxis' as well as other forms of public transportation (and vehicles in general) make frequent stops. Blanchisseuse Road north of Arima is a winding, narrow, paved path that snakes its way through the Arima Valley. There are many areas with fairly steep drop-offs. I strongly suggest honking one's horn at blind curves, and driving the road during daylight first.

Forest Type

The natural vegetation in the area of the Asa Wright Nature Centre and Lodge is classified as Tropical Wet Forest.

Much of the vegetation on the AWNCL land itself is secondary growth with agricultural remnants (cacao, citrus, etc.).

Seasonality

The main dry season occurs approximately from January to late May or early June, with a small dry period (*petit carême*) also occurring in October. The wet season normally extends from June through December. Average rainfall has been recorded at approximately 3,000 mm annually.

Facilities

The Asa Wright Nature Centre and Lodge has 200 acres of land, previously used as an agricultural estate. The main building was an 80 year old estate house which housed the reception lobby, administrative office, six guest rooms, dining room, kitchen, sitting room, and large elevated porch for observing birds. In the reception lobby were conservation displays and showcases with articles and books written about various aspects of the local flora and fauna. There were also a variety of books on birds, flowers, trees, and insects of the Caribbean for sale.

Rear view showing the screened porch attached to the back of the estate house at the Asa Wright Nature Centre and Lodge. (Near Arima, Trinidad)

The end of the lobby led to the dining room. Here, several round tables were arranged to accommodate 8-10 people each. Large 'lazy-susans' occupied the middle of each table which were decorated with local tropical flowers. I did not stay at the AWNCL as a guest and cannot comment on the food personally, but the several guests I questioned all told me it was excellent.

A hallway led from the center of the lobby, past two guest rooms and to a sitting room. This room appeared to have antique furniture which may be from the original estate. The walls were decorated with many beautiful prints of tropical birds. A glass case along one wall held an excellent library, featuring numerous bird volumes, but also many other books, journals, and magazines of interest to the naturalist in general.

The sitting room exited to a large (approximately 7m x 20m) screened back porch that overlooked the tree-filled property and adjacent mountains. Here one could sit and relax and observe the almost constant bird activity at the strategically placed hummingbird feeders and fruit feeding stations. The flowering and fruiting trees surrounding the house also attracted naturally many tropical birds. The porch was comfortably furnished with chairs, couches, and tables and has additional bird prints and pictures on the wall. The porch windows slid to give an unobstructed view to birders and photographers. There was also a VCR and television monitor with a series of special natural history tapes available to the visitor.

There were two guest rooms on the main floor of the house, used primarily for couples. They contained a single large bed, armoir or closet and shelves, electric lights and outlets for 115V electricity, and a private bathroom with sink, flush toilet, and hot water shower. The water at the AWNCL was potable and came from their own spring. Towels and linens were provided. There were an additional four rooms in the lower level of the house, but these may be changed in the near future to make room for a small museum. During my visit, each of these lower rooms accommodated two people and had the same amenities as mentioned previously, but with two beds.

Additional lodging facilities then available at the AWNCL were two cottages that held four people each, a bungalow for 6-8 people, another separate building with two rooms for four people total, and a final separate building with three rooms used as singles. This put the total capacity at approximately 35 persons. It was expected however, that lodging for 44 persons would be available by October 1989.

Other attractions on the property were a covered barbecue area, often used by visiting groups and day visitors. There was also a waterfall and natural pool accessible on the property.

Trail System

The Asa Wright Nature Centre and Lodge had a network of several kilometers of trails that are usually marked or identified at their beginning. Excellent trail maps were available to guests and visitors. Several of the trails cut through the display grounds and territories of specific bird species such as the white-bearded manakin, the golden-headed manakin, and the bellbird. Such locations are indicated on the trail map. The degree of difficulty and time needed to walk each trail was also described on the trail map and pamphlet.

The AWNCL also provided guides and guided excursions to ecological points of interest on and off the Centre's property. These were half- or full-day field trips and could include a visit to Dunston's Cave to see the oilbird (*guacharo*) colony, a trip to Caroni Swamp, a climb in the forest of the Aripo Valley, or a nocturnal foray to see nesting sea turtles (in season), among others.

Costs

Based on information I received in November 1989, the cost for single accommodations at the Asa Wright Nature Centre and Lodge was $77 (from May 1 - November 30) and $98 (from December 1 - April 30). The cost per person based on double occupancy was $56 (from May 1 - November 30) and $72 (from December 1 - April 30). These rates included lodging, three meals a day, an afternoon tea, and a rum punch each evening. All rates were subject to a 10% service charge. Roundtrip transfer from Piarco Airport to the AWNCL cost $35. Most half day field trips were $15 per person and full day field trips were $30 per person. There was a $5 daily fee for non-guest visitors who wished to used the AWNCL grounds.

Comments

My exposure to the AWNCL was based, unfortunately, only on several short visits from the nearby Simla Research Station where I was staying. However, each visit left the impression of an efficient, well-run operation where guests are catered to in traditional British manner. The guests I spoke with had no complaints and seemed well satisfied with both the service and accommodations they received.

The AWNCL appears to be a comfortable center from which birdwatcher, tourist, or naturalist can explore the Arima Valley and other parts of the island. Biologists wishing to conduct research in the area are advised to work from the Simla Research Station (see next section).

Simla Research Station
(William Beebe Tropical Research Station)

Contact

    Asa Wright Nature Centre and Lodge

Address

    The Manager
    Asa Wright Nature Centre
    P.O. Bag 10
    Port-of-Spain
    Trinidad, West Indies

    Port-of-Spain Office
    Telephone: (809)-667-4655

    U.S. Representative
    Caligo Ventures
    387 Main Street
    P.O. Box 21
    Armonk, NY 10504-0021 USA
    Telephone: (914)-273-6333 (Local)
    (800)-426-7781 (Toll Free)
    (800)-327-2753 (In New York)

Description

Simla was originally a country estate with a house built around 1920 for the Governor General of the island. In 1949, William Beebe purchased the 22-acre estate and building and presented it to the New York Zoological Society to serve as the headquarters for their Department of Tropical Research, which he eventually directed. From 1950-1970 much research was carried out at Simla and additional lands were added to the estate so that it now encompasses 265 acres. After Beebe's death in 1962, Simla became known as the William Beebe Tropical Research Station. The station eventually closed and was incorporated into the Asa Wright Nature Centre in 1974.

Today Simla is only available to researchers. Permission to visit or conduct research at Simla can be obtained from the Asa Wright Nature Centre and Lodge.

Location

The Simla Research Station is located on the eastern side of the Arima Valley in Trinidad's Northern Range. It is situated approximately 6.5km north of the north central city of Arima. The island of Trinidad lies between 10° and 11° North latitude, with Simla located at an elevation of approximately 265 meters.

Logistics

The Simla Research Station is about 40 km from Port-of-Spain and 18 km from Piarco Airport. Either the Eastern Main Road or Churchill Roosevelt Highway can be used to travel east from Port-of-Spain or the airport. They can be reached by exiting Port-of-Spain in the east, and when coming from Piarco Airport by first traveling north on Golden Grove Road. The Eastern Main Road is followed directly into the town of Arima. Travelers on Churchill Roosevelt Highway should look for signs for Arima and take either O'Meara Road or Tumpuna Road north into the town itself.

Once in Arima, one must find Blanchisseuse Road which will be followed north and uphill. The entrance road is approximately 100 meters past the 4-mile post (a white cement post with a black 4 on it) above the town of Arima. There was a battered green sign indicating a quarry where visitors need to turn right off of Blanchisseuse road to reach Simla. This road is followed for approximately a kilometer until another road is seen on the left. This left hand paved road is the entrance drive to Simla and should be blocked by a lockable, raisable metal bar type barrier (if locked, the key can be obtained from the Asa Wright Nature Centre and Lodge [see pages 146-151]). At the point where one turns left into the Simla entrance drive, there should be a shack on the right hand side, and the way ahead should also be blocked by a barrier gate belonging to the quarry. The Simla entrance is a narrow, paved way which extends for slightly less than a kilometer, terminating in a grassy area with a covered car port adjacent to the field station.

Forest Type

The area near the Simla Research Station is classified as Tropical Wet Forest, although other ecological habitats on the island are easily accessible.

Seasonality

The main dry season at Simla normally extends from January to late May or early June, with a small dry period (*petit carême*) during October. The rainy season usually occurs from June through December. Annual rainfall averages about 2,750mm.

Facilities

The facilities at Simla were housed in three buildings situated in a clearing of the forest atop a ridge. The main building was a 72 year old house that used to belong to the Governor General of the island. This house contained five rooms (approximately 3.5m x 4m each) for lodging visitors. The typical room had two double bunk beds, enabling four persons to share each room. However, unless filled to capacity, visitors were given separate rooms. Each room was screened with large windows with wooden shutters. Each room had a sink with cold, potable water piped in from a mountain stream. Towels, linens, and soap were provided. Rooms also had shelves and hangers for clothes or an armoire. Simla was provided with 115V electricity and each room had electrical outlets that take the now standard plug with two parallel, flat, metal prongs, one side wider than the other.

There were two separate bathrooms and two separate showers that were shared by the guests in the field station building. The bathrooms contained one sink and one flush toilet each. The showers had 24 hour hot water.

The inner kitchen contained a refrigerator with freezer, a sink with hot and cold running potable water, a large counter and cabinets for foodstuffs, a toaster, and pots, pans, utensils and other cooking necessities. The outer kitchen was housed in a small, detached room about three meters away. It was reached by crossing a short, covered courtyard area. The outer kitchen contained a four burner gas range with oven, a stand up freezer, and a small table. The dining room was approximately 5m x 5m and contained a long table capable of seating 12 people. A coffee table and chairs were also present.

The sitting room took up one end of the building and measured approximately 5m x 10m. It had couches, several chairs, two desks and could easily be used for lectures, seminars, or slide shows. Folding doors at one end opened up to give an excellent 'open wall' view of the Arima Valley (facing north). There was a small administrative office with some books, reprints, and scientific papers at the opposite end of the sitting room.

Four laboratories composed the remaining inside rooms. They measured approximately 2m x 5m, 3m x 5m, 3m x 4m, and 3.5m x 4.5m. All had windows and a sink with running water; one also had two large 'dry boxes' for moisture sensitive items. Along the south side of the building was a large open cement patio, where outdoor experiments could be set up and conducted.

Immediately outside of the main building, near the outer kitchen, was a small covered room containing a scrub sink, electric washing machine, and electric dryer. Both appliances functioned well during my visit.

Approximately 25 meters (east) behind the main building were two newer, smaller buildings. One is a bungalow containing five rooms for guests. Although I did not have the opportunity

Main building of the Simla Research Station (formerly the William Beebe Tropical Resesarch Station). (Near Arima, Trinidad)

to examine any of these rooms, I assumed they were furnished similarly to the other visitors' rooms. The second building was actually a small cottage that is also available for rent to researchers. This small, dual-level house was somewhat shabby, but could easily be fixed up. The first floor had two main rooms (approximately 4m x 8m and 4m x 6m) set up with tables, chairs, counter space, and makeshift desks. A refrigerator was in one corner. A small kitchen area was located in one end of the smaller room. It contained a gas range with four burners, sink, counter, cabinets, shelves, and dishes. There was a third, irregularly shaped side room (approximately 2.5m x 3m) on the first floor.

A stairway led down to the bottom floor and sleeping quarters. There was one large bedroom (approximately 4m x 6m) with two single beds, night table, armoir, and dresser. The remaining room was the bathroom with sink, flush toilet, and hot water shower. There was electricity (110V) throughout the house and each room had overhead electric lights.

At the Simla Research Station, visitors must bring and prepare their own food. Supplies can be purchased in the nearby town of Arima, approximately 20 minutes away by car.

Trail System

There are about 265 acres of land on the Simla estate, with several trails that lead off from the field station. These trails are neither marked nor mapped. In addition, there are also rustic, unpaved roads called 'traces' which can be walked through forested habitat. Visitors to the Simla Research Station are also close to the Asa Wright Nature Centre and Lodge, which has an excellent set of mapped trails (see page 150).

Costs

During June 1989 the charge to researchers staying at the Simla Research Station was $10 per day per person for the first 21 days, and then $7.50 per day thereafter. To rent out the cottage, one must consult the management for prices. The station fees were paid at the administrative office of the Asa Wright Nature Centre and Lodge (see pages 146-151).

Comments

The Simla Research Station is a facility with enormous potential. It can function as a headquarters for researchers or serve as a comfortable location to conduct classes and seminars related to tropical biology. Conditions of the field station have greatly improved since I visited there in 1985. Holes in screens have been repaired, the washer and dryer now both function, and the electricity appears to have been renovated. The live-in manager, Francis Morean, is very knowledgeable about local flora and fauna and has always been most helpful during my visits. Although I have never had any problems with theft, security at Simla is minimal and there are often times when the building has been completely open and unoccupied for hours. One biologist I spoke with who stayed at the field station for several months did report having repeated problems with theft of parts from his car. The most negative characteristic of Simla is the noise and visual pollution caused by the nearby gravel quarry.

The Simla Research Station appears to suffer from lack of funds and indecision by its board of directors, who are primarily concerned with the management and administration of the Asa

Wright Nature Centre and Lodge. However, Simla continues the tradition of tropical biology research in Trinidad started by William Beebe and the New York Zoological Society. Due to its location on a small island such as Trinidad, one can easily drive to a variety of different ecological habitats. Easy access to the South American mainland is also provided through the international airlines serving Piarco Airport.

Victoria Regia Research Station

A new research station (Victoria Regia) opened in central Trinidad in late 1988. It is open to scientists, students, birdwatchers, and tourists. For additional information write to the following address.

> Jack and Caroline Price
> Victoria Regia Research Station
> La Gloria Road
> Talparo, Mundo Nuevo
> Trinidad, West Indies

Comments On Trinidad

Trinidad is a unique and unusual island. It shares many species in common with the South American mainland, which is only eight miles away. Its small size makes it possible for one to travel throughout the island in a limited amount of time. The Asa Wright Nature Centre and Lodge and Simla provide two very different sets of facilities for visiting tourists and biologists, respectively. The British influence can still be seen on the island (as in left-side driving), while the people form a distinctive ethnic and cultural mixture atypical of other Caribbean islands.

Books

CARIBBEAN FLORA
Adams, C. Dennis
1976. Nelson Caribbean

POPULAR TOURS AROUND TRINIDAD
Bacchus-Xavier, Joan
Educo Press Publishers (Barataria, Trinidad)

FLORA AND FAUNA OF THE CARIBBEAN - AN INTRODUCTION TO THE ECOLOGY OF THE WEST INDIES
Bacon, Peter R.
1978. Key Caribbean Publications

BUTTERFLIES OF TRINIDAD AND TOBAGO
Barcant, Malcolm
1970. Collins Clear-Type Press

A GUIDE TO THE BIRDS OF TRINIDAD AND TOBAGO
ffrench, Richard
1976. Harrowood

NATURE TRAILS OF TRINIDAD AND TOBAGO
ffrench, Richard and Peter Bacon
1982. S.M. Publications

TROPICAL BLOSSOMS OF THE CARIBBEAN
Hargreaves, Dorothy and Bob Hargreaves
1960. Hargreaves Company, Inc.

TROPICAL TREES - FOUND IN THE CARIBBEAN, SOUTH AMERICA, CENTRAL AMERICA, AND MEXICO
Hargreaves, Dorothy and Bob Hargreaves
1965. Hargreaves Company, Inc.

THE BIRDS OF TRINIDAD AND TOBAGO
Herklots, G.A.C.
1961. Collins

SITES (TRAIL, GUIDE) - TRINIDAD AND TOBAGO
Jackson (Langton), Sarah (Producer)
1986. Trinidad and Tobago Tourist Board

NATIVE ORCHIDS OF THE EASTERN CARIBBEAN
Kenny, Julian
1988. Macmillan Caribbean

FLOWERS OF THE CARIBBEAN
Lennox, G.W. and S.A. Seddon
1978. Macmillan Caribbean

A BIRDER'S GUIDE TO TRINIDAD AND TOBAGO
Murphy, William L.
1986. Peregrine Enterprises, Inc.

GUIDE TO THE IDENTIFICATION OF THE AMPHIBIANS AND REPTILES OF THE WEST INDIES (EXCLUSIVE OF HISPANIOLA)
Schwartz, Albert and Robert W. Henderson.
1985. Milwaukee Public Museum

A NATURALIST IN TRINIDAD
Worth, C.B.
1967.

Maps

Decent road maps of the island were not easily available in major bookstores, either in Port-of-Spain or at the University of the West Indies. The only place with detailed road maps for sale during my visit was the office of the Trinidad and Tobago Tourist Board at:

    CIC Building
    122/124 Frederick Street
    Port-of-Spain
    Trinidad, West Indies
    Telephone: 623-1142

Detailed physical-political maps of Trinidad can be obtained at the Red House (a government building) in Port-of-Spain. It is located on Abercromby Street, between Knox and Hart Streets across from Woodford Square.

For domestic sources of Latin American and Caribbean maps see pages 290 and 291.

Tourist Information Sources

Maps, books, and general information can be obtained from the Trinidad and Tobago Tourist Board, whose addresses are listed below.

Trinidad
CIC Building
122/124 Frederick Street
Port-of-Spain
Telephone: 623-1142

New York
15th Floor, Forest Hills Tower
118-35 Queens Boulevard
Forest Hills
New York, NY 11375
Telephone: (718)-575-3909

Miami
330 Biscayne Boulevard
Suite 310
Miami, FL 33132
Telephone: (305)-374-2056

Canada
40 Holly Street, Suite 102
Toronto
Ontario M4S 3C3
Telephone: (416)-486-4470

United Kingdom
48 Leicester Square
London WC2H 7LT
Telephone: (01)-930-6566

Trinidad and Tobago Hotel and Tourism Association
Trinidad Hilton
POB 243
Port-of-Spain
Trinidad, West Indies
Telephone: 624-3065   Telex: 22319

Scientific Organizations/Institutions

Trinidad and Tobago Field Naturalists' Club
c/o Honorary Secretary
1 Errol Park Road
St. Ann's
Trinidad, West Indies

University of the West Indies
Dept. of Zoology
St. Augustine
Republic of Trinidad and Tobago
West Indies
Telephone: 66-31364 or 66-31678

Commonwealth Institute of Biological Control
Gordon Street
Curepe
Trinidad, West Indies
Telephone: 66-24173

CHAPTER 1F   COSTA RICA

General Information

Area: 51,000 sq km (19,600 sq miles)

Population: 2,922,000

Capital: San Jose (population: 890,000)

Language: Spanish

Currency: Colones (1 colón = 100 céntimos)
            Coins: 25 and 50 centimos; 1, 2, 5, 10, and 20 colones.
            Notes: 10, 50, 100, 500, 1,000, and 5,000 colones.
            Exchange Rate: U.S. $1 = 76.1 colones (August 1988)
                           U.S. $1 = 79.4 colones (January 1990)

National
Airlines: LACSA, SANSA

Telephone
Code: Costa Rica: 506
            No special city codes needed

Costa Rica

1 - Monteverde Cloud Forest Reserve 2 - La Selva Biological Station
3 - Selva Verde Lodge 4 - Green Turtle Research Station
5 - Marenco Biological Station

Introduction

Costa Rica is the second smallest Central American country, approximately a little less in size than the state of West Virginia. It shares its northern border with Nicaragua and its southern border with Panama, while its east coast lies along the Caribbean Sea and its west coast along the Pacific Ocean. Numerous mountain ranges extend in a chain from the northwest to the southeast through the central portion of the country. There are coastal plains on either side of these mountains, the largest being to the north and east of the chain. Two sizable peninsulas jut out into the Pacific Ocean from Costa Rica's southern coast. The larger is the Nicoya peninsula, located near the northwest corner of the country and forming a gulf of the same name (Golfo de Nicoya). The smaller or Osa Peninsula exists near the southeast corner of the country and forms the Golfo Dulce.

Costa Rica has no standing army and a long history of political stability. Its citizens (called Ticos) enjoy the highest standard of living in Central America. Like most neotropical countries, much of Costa Rica's forests are being destroyed or converted into farm and pasture land. However, unlike most, a large amount (approximately 20% of the total land area) is protected as a national park or nature reserve. This conservation-oriented tradition, combined with the great diversity of habitats and ecosystems found throughout the country, have made Costa Rica a haven for both tourists and biologists. World reknown scientists and conservation leaders such as Archie Carr and Dan Janzen have spent decades working in Costa Rica. Today, many students of tropical biology from all over the world receive their initial field training in Costa Rica.

My evaluation of rainforest sites in Costa Rica took place during July-August 1988. At this time I visited three separate locations: two in the lowlands of northeastern Costa Rica (La Selva Biological Station and the Selva Verde Lodge), and one in the mountains of the northwest (the Monteverde Cloud Forest Reserve). Information is also provided on the Marenco Biological Station, the Wilson Botanical Garden, and the Green Turtle Research Station.

Monteverde Cloud Forest Reserve

Contact

    Tropical Science Center

Address

    Apartado 8-3870
    San José, Costa Rica
    Central America
    Telephone: 22-62-41

Description

    The Monteverde Cloud Forest Reserve is a privately owned biological reserve that acts as a cooperating unit with the Costa Rican National Parks System. It was established in 1972 through private donations by conservation organizations and individuals. It is administered by the Tropical Science Center, a San Jose-based non-profit conservation organization.

Location

    The cloud forest reserve is part of Monteverde, a town located approximately 92km northwest of San Jose. It is situated in the province of Puntarenas in the mountains of the Cordillera de Tilaran, near the point where the neighboring provinces of Guanacaste and Alajuela meet. Monteverde is situated between $10°$-$11°$ North latitude.

Logistics

    Public transportation to Monteverde is easily accessible. In San Jose, one can catch a bus that goes directly to Monteverde at the *parada* (bus stop) located on Avenida 9, between Calles 10 and 12. The cost was approximately 230 colones ($3) and the ticket was purchased on the bus after the trip had started. Heavier baggage was stored in a compartment in the rear of the bus, from the outside. The ride took approximately 3 1/2 - 4 hours (if no delays), with a stop of about 20 minutes at a restaurant approximately half way there. This direct bus left on Monday, Tuesday, and Thursday at 2:30 pm; and on Saturday at 6:30 am. There were several small stores by the bus stop area in San Jose where sandwiches, coffee, soda, etc. can be purchased.

The direct bus passed through the small town of Santa Elena, located about 3km away from Monteverde. The bus went as far as the cheese factory (*la fábrica que produce queso*), which is the most prominent landmark in the community of Monteverde. Passengers can be let out in Santa Elena, or further along towards the cheese factory, depending on where they are staying. The direct bus left Monteverde for San Jose from the cheese factory on Sunday and Friday at 3:00 pm, and on Tuesday and Thursday at 6:30 am.

Two other methods of reaching Monteverde by bus were to take the morning bus from San Jose to Puntarenas, and then the afternoon bus from Puntarenas to Santa Elena. Or one could take the bus to Tilaran from San Jose and get off at the Rio Lagarto, and catch the bus for Santa Elena when it passed by. Taxi fare from Santa Elena to Monteverde was about 300 colones ($4).

To reach Monteverde by car, follow the Pan American Highway north from San Jose to Rio Lagarto. Turn right just before the river and continue for approximately another 40km.

Forest Type

Tropical Cloud Forest (including eight different ecological life zones), is present at the Monteverde Cloud Forest Reserve.

Seasonality

The dry season at the Monteverde Cloud Forest Reserve usually begins around mid-December and continues through to the month of May, with February and March being the driest months. The rainy season starts in late May and runs through to early December, with October and November normally being the wettest months. There is a short dry period of 2-3 weeks that usually occurs in July. A 31-year average of the annual rainfall in the Monteverde area from 1956-86 is 2,518mm.

Facilities

The Monteverde Cloud Forest Reserve presently encompasses an area of approximately 10,000 hectares, including a portion initially set aside by the original Quaker inhabitants of the Monteverde community to protect their watershed. The Reserve occupies land on both the Caribbean and Pacific slopes of the continental divide in the Cordillera de Tilaran. Elevations range from 800 - 1,800 meters and include a wide diversity of ecological habitats.

The field station is available to scientists and other visitors through arrangements made with the Tropical Science Center. Available at the field station are cooking facilities, a simple laboratory, and dormitory style housing for up to 30 people. There is also a campground near the field station where campsites can be rented for a nominal fee.

Field station maintained by the Tropical Science Center at the Monteverde Cloud Forest Reserve. (Near Monteverde, Costa Rica)

Trail System

There is an excellent series of clearly marked trails that extend throughout the Reserve. Trail maps showing their layout, points of interest, and the location of shelters can be obtained by visitors at the visitor center/field station, located at the Reserve entrance.

Costs

I stayed at the Pension Flor Mar in Monteverde, located approximately 2km from the Reserve's field station at the base of the entrance road. There are several very inexpensive *pensiones* (small hotels or rooming houses) in Santa Elena, although the *pensiones* closer to the reserve are still not expensive by North American standards. The Flor Mar cost approximately $18/day, including all three meals. Patrons are asked at breakfast if they would prefer a bagged lunch to take with them into the Reserve. There are seven rooms in the Flor Mar, with an additional three in the owner's adjacent house, for a total of 35 beds. The bathrooms have showers with hot water and flush toilets, although few rooms have a private bath included.

The Pension Flor Mar, a small rooming house approximately 2km from the Monteverde Cloud Forest Reserve. (Monteverde, Costa Rica)

The Flor Mar is owned and operated by Marvin Rockwell and his wife Flory. Mr. Rockwell is one of the original Quakers who arrived in Costa Rica in 1951 and helped establish Monteverde. He has acted as a tour guide to sites throughout Costa Rica, and can give much valuable information and advice about traveling in the country. The Pension Flor Mar is located only a short walk (unless one is loaded with equipment) up the road beyond the cheese factory. It is the closest *pensión* to the Monteverde Cloud Forest Reserve. The address is:

    Pensión Flor Mar
    Apartado 10165
    San José, Costa Rica
    Telephone: 61-09-09

There is a daily entrance fee of approximately $3 to enter the Monteverde Cloud Forest Reserve. This is paid at the visitor center, where gifts and booklets about the Reserve's flora and fauna can also be purchased.

Comments

Scientists have inventoried over 2,500 species of plants, 400 species of birds, and 100 species of mammals from the Reserve. The Monteverde Cloud Forest Reserve is the only known area that supports the golden toad, and is one of the few remaining refuges of the resplendant quetzal. Due to the Reserve's location and geography, many of the trees are covered with moss and epiphytes.

Monteverde is easily accessible by car or bus and well worth the trip. The community has a rich and interesting history, having been founded by Quakers from Alabama during the 1950s. The town is well known in Costa Rica for the cheese it produces, but only in the past ten years has it gained equal recognition for its cloud forest reserve. Many of Monteverde's inhabitants speak English and are accustomed to the many tourists and biologists that visit the area. Visitors to Monteverde should be aware that although they are in the tropics, due to the altitude the weather may be cool. A sweatshirt or jacket is recommended. During my stay biting insects were not a problem. Food items can be purchased in Santa Elena or at the cheese factory in Monteverde.

Efforts are underway by the Monteverde Conservation League (a non-profit fund-raising organization) to raise money to purchase additional land at lower altitudes, adjacent to the Reserve. Many of the animal species (like the quetzal) are altitudinal migrants, spending a part of the year at lower elevations. Their protection can only be assured with the acquisition and management of land areas at lower altitudes. For more information contact the following agencies:

Monteverde Conservation League
Apartado 10165-1000
San José, Costa Rica
Telephone: 61-09-54 or 61-09-55

World Wildlife Fund
International Program: MONTEVERDE
1255 23rd Street, NW
Washington, D.C. 20009 USA

Nature Conservancy
International Program: MONTEVERDE
1785 Massachusetts Ave., NW
Washington, D.C. 20036 USA

La Selva Biological Station

Contact

> Organization For Tropical Studies

Address

> North American Office
> Box DM
> Duke Station
> Durham, N.C. 27706 USA
> Telephone: (919)-684-5774

Contact

> Organización Para Estudios Tropicales

Address

Oficina Centroamericana
Apartado 16
Universidad de Costa Rica
San José, Costa Rica
Telephone: 36-66-96

Description

The La Selva Biological Station is owned and operated by the Organization For Tropical Studies (OTS). Its purpose is to foster education and facilitate field research in tropical biology.

Location

The La Selva Biological Station (also known as Finca La Selva, and to be referred to for the rest of this section simply as La Selva) is located approximately 55km north and slightly east of San Jose, when measured on the map. It is situated in the Atlantic lowlands of Heredia Province, near the town of Puerto Viejo at the confluence of the Rio Sarapiqui and Rio Puerto Viejo. La Selva falls between 10°-11° North latitude.

Logistics

The Organization For Tropical Studies' Central American Office is located about 10-15 minutes by car (in light traffic) from central San Jose. It is situated approximately 450 meters west of the Colegio Lincoln in the area known as Moravia, near the border of Tibas. An OTS vehichle makes passenger trips from the OTS office to La Selva and then back again on Mondays, Wednesdays, and Fridays. The car makes a direct trip, leaving the OTS office at 8:00am, and then leaving La Selva for San Jose at 1:00pm. The trip takes from $2^{1}/_{2}$ - $3^{1}/_{2}$ hours depending on the route and weather conditions, as one must travel approximately 100km by road. The cost is $10 (one way) for senior researchers and $6 for graduate students without support. Reservation for space on the OTS vehichle should be placed at least 24 hours in advance by calling the OTS office in San Jose.

For those wishing to use public transportation, there are two to three buses that run daily between San Jose and Puerto Viejo. These buses start from the Puntarenas bus stop on Avenida 9, between Calle 10 and Calle 12. The trip usually takes from $3\,^1/_2$ - 5 hours. There is also transportation (daily, except Sundays) from Puerto Viejo-La Selva-Puerto Viejo. One should contact the OTS San Jose office or La Selva for the current bus and transportation schedules.

Visitors driving their own vehichles can obtain a map of the route to La Selva from the San Jose office. Basically, there are two main ways. One can take the Interamerican Highway north to Heredia and then follow Route 9 to Puerto Viejo, passing on the way Carrizal, Vara Blanca, La Virgen, and Chilamate. Or one can take the road that begins between San Jose and Heredia and follow it northeast. This will take you through a portion of the impressive Braulio Carillo National Park, past Santa Clara and Rio Frio, where the road bends westward. One continues on past Tigre, and La Selva is shortly beyond.

Forest Type

The forest at La Selva is classified as Premontane Tropical Wet Forest.

Seasonality

There is an annual precipitation of approximately 4,000mm, with 100mm or more every month. The drier period occurs between January and May.

Facilities

The La Selva Biological Station contains more than 1,500 hectares (3,700 acres) of land, including various types of ecosystems. It adjoins Braulio Carillo National Park which contains 45,000 hectares of protected land. The elevation range of La Selva is 35-150 meters. It includes primary forests, secondary forests, cacao plantations, pastures, swamps, rivers, and creeks. Approximately 90% of the original La Selva reserve is undisturbed

Dining hall and administrative offices of the La Selva Field Station operated by the Organization for Tropical Studies. (Near Puerto Viejo, Costa Rica)

Newly constructed modern laboratory at the OTS La Selva Field Station. (Near Puerto Viejo, Costa Rica)

Long-term researcher's cabin at the OTS La Selva Field Station. (Near Puerto Viejo, Costa Rica)

Student or field class bungalow at the OTS La Selva Field Station. (Near Puerto Viejo, Costa Rica)

tropical rainforest. Its flora is estimated at greater than 2,000 species, while records for other groups are as follows: 407 birds, 143 butterflies, 63 bats, 122 reptiles and amphibians, and 42 fish.

La Selva has extensive research facilities available to visiting biologists. A complete description can be obtained from either the North American or Central American offices of OTS. There are two large modern laboratories(5,000 and 4,000 sq ft), 24 hour electricity, a library, a herbarium and reference collections, shade houses, air-conditioned work space for 30 people, screened work space for 14 people, and a screened seminar/class room that holds 60 people. There is also telephone service available at the station, permitting overseas calls.

Included among the laboratory equipment are the following: deionized water still, fumehoods, vacuum pumps, refrigerators, chromatography refrigerator, freezers, ice machine, dry ice machine, drying ovens, analytical balances, top-loading balances, table top and high speed centrifuges, pH meter, shaker table, vortex mixer, ultrasonic cleaner, rotary evaporator, freeze dryer, incubator, water baths, autoclave, oscilloscope, microosmometer, PMS pressure chamber, porometer, Heinz Waltz portable leaf fluorometer, oxygen temperature-meter probe, Licor leaf-area meter, Licor 1800/22 spectroradiometer, spectrophotometers (Spec. 20 and Perkin Elmer 3A UV-Vis), Decagon photosynthesis system, gas chromatographs (Shimadzu 9 AM with FID, ECD, TCD and integrator), chart recorders, desiccators, plant tissue grinders, soil grinder, Wiley mill, autoanalyzer, root elutriator, digestion block, muffle furnace, digital current meter, digital fluorometer, underwater photometer, dissecting and compound microscopes (brightfield and UV), microcomputers (IBM compatible). Shop facilities and a 4-wheel drive vehicle are available for local rental.

Living accommodations are very comfortable, ranging from dormitories to shared or private cabins, depending on one's length of stay and the number of people at the station. There is a staff of full-time cooks that serve three meals a day in a modern dining hall/cafeteria, which is always open for tea, coffee, etc. Washers and dryers are also available for the use of station visitors.

Trail System

La Selva has an extensive, well-marked trail system. The station grounds are plotted with a 200 meter grid system, with numbered posts marking all intersections. Trails are marked every 50 meters with the distance from the starting point. Trails are easy to follow, and trail maps are available at the station and OTS offices. A large arboretum is located at La Selva, containing more than 1,300 tree specimens.

Costs

Accommodations at La Selva include room and board, the use of all facilities, and access to the forest. A full schedule of rates can be obtained from either of the OTS offices. Rates in effect as of 1 January 1989:

Senior Researcher	$35/day or $210/week
Graduate Researcher	$22/day or $130/week
Visitor/Tour Group	$70/day

Permits To Conduct Research

All research conducted at La Selva must be approved by OTS. Investigators should submit a research proposal that describes the nature of the study, duration, methods, number of participants, source of funding, and dates for which the work is proposed. Scientific justification must be given for any proposed collecting on the La Selva grounds. If materials are to be collected and exported, the researcher must obtain the appropriate Costa Rican permits. The procedures for obtaining such permits can be provided by the OTS office in San Jose, which suggests allowing at least one day at the beginning and end of your trip to accomplish the necessary paperwork. Copies of research proposals must be submitted at least two months in advance of the proposed starting date. Proposals should be sent to both:

Dr. Donald Stone
Executive Director
Organization For Tropical Studies
P.O. Box DM
Duke Station
Durham, N.C. 27706 USA

And

Drs. David and Deborah Clark
Co-Director
La Selva Biological Station
Organization For Tropical Studies
Apartado 676
2050 San Pedro de Montes de Oca
Costa Rica, America Central

Comments

The La Selva Biological Station is probably the most luxurious and comfortable biological research station in the Neotropics. It is extremely well run and organized, with OTS having done everything possible to encourage and facilitate long-term research. It is easily accessible from San Jose by a paved road that runs very close to the actual La Selva property. The extensive, marked trail system provides easy access to the many different types of habitats with little fear of getting lost.

Although tourists are permitted to stay at La Selva, it is a facility primarily for biologists and researchers, who are given preference. An agreement has recently been reached which will allow the guests of the nearby Selva Verde Lodge (see next section) to visit the La Selva grounds and receive a supervised tour for a fee.

Selva Verde Lodge

Contact

Giovanna and Juan Holbrook

Address

Holbrook Travel Agency
3540 NW 13th Street
Gainesville, FL 32609 USA
Telephone: (904)-377-7111 or (800)-451-7111 (USA) or
(800)-341-7111 (Florida)
Telex: TWX 810-825-6340 Fax: (904)-371-3710

COSTA RICA

In Costa Rica contact:

Selva Verde Lodge
Chilamate, Puerto Viejo
Costa Rica, Central America
Telephone: 71-64-59

Description

The Selva Verde Lodge is a tourist facility that provides access to much of the same vegetation, flora, and fauna that is found at the nearby La Selva Biological Station (see pages 171-178).

Location

The Selva Verde Lodge is located in the village of Chilamate (approximately 70 inhabitants), in the Atlantic lowlands of Heredia Province. Chilamate is situated between the towns of Puerto Viejo and La Virgen, about 30km due north of San Jose. Selva Verde is approximately 12km (a 15 minute drive) from the La Selva Biological Station, and about 18km from Braulio Carillo National Park. It is located between 10°-11° North latitude.

Logistics

Selva Verde is approximately a 2½ hour drive from San Jose, when taking the paved highway (Route 9). The lodge and adjacent restaurant are situated along side the road and are clearly marked by a colorful hanging sign that reads Finca Selva Verde - - Albergue y Restaurante. Those wishing to use public transportation from San Jose can use the buses destined for Puerto Viejo (see previous section on the Logistics of the La Selva Biological Station).

Forest Type

Premontane Tropical Wet Forest is the classification of the forest vegetation at Selva Verde.

Seasonality

The drier period occurs from January to May. Precipitation reaches at least 100mm each month and approximately 4,000mm annually.

Facilities

The Selva Verde Lodge was founded in 1985 and encompasses 500 acres of land. Approximately 20 acres of this land is located between the road bordering the lodge and the Rio Sarapiqui. It is in this area that the lodge facilities and restaurant were located. The remaining land has been maintained as a preserve and consists primarily of undisturbed rainforest. There was no marked trail system, but a guide is available at a nominal rate to lead guests on excursions into the forest.

At the time of my visit, the Selva Verde Lodge consisted of 10 guest rooms able to accomodate approximately 22 people. All rooms were screened and had an electric fan. There are seven showers and toilets that were shared by the guests, located at one side of the lodge. Additional facilities were under construction. These have since been completed and provide 24 double guest rooms with private baths. Adjacent to the lodge was a screened-in, thatched-roof restaurant where guests took their meals.

There was 110V electricity at the lodge and outlets that accepted the standard plug with two flat, parallel prongs.

Costs

The following are 1990 rates for the Selva Verde Lodge, obtained from the Holbrook Travel Agency.

	Single	Double	Triple
Creek Lodge	$52	$42	$37
River Lodge	$57	$49	$42

The above rates are per person, per day. Children under 12 are $25 per day. These prices include breakfast, lunch, and

Rear view of the Selva Verde Lodge. (Chilamate, Costa Rica)

dinner. Meals are Costa Rican cuisine and served buffet style. Box lunches to take on excursions are available by requesting them the night before. Prices include a 13.3% tax, but do not include gratuities.

Comments

The Selva Verde Lodge appeared to be a clean, quiet, well run lodge where travelers can enjoy the beauty of the Costa Rican countryside and wildlife. Although rustic in appearance, both the lodge and the restaurant were attractively built and afforded comfortable accommodations. It is highly recommended for visitors who wish to see the same type of forest habitat as La Selva, but who are not biologists or do not wish to be in a research-oriented atmosphere. Guests of the Selva Verde Lodge are permitted to make supervised visits to nearby La Selva, if prior arrangements have been made.

The Selva Verde grounds near the lodge were landscaped with tropical plants to attract butterflies and hummingbirds. Guests at the lodge had access to a large selection of books on Costa Rica and its wildlife.

Rara Avis

Amos Bien
Rara Avis S.A.
Apartado 8105
San José 1000
Costa Rica
Telephone: 53-08-44 and 24-99-56

Rara Avis is a 1,500 acre rainforest site with facilities for visitors located approximately 40km due north of San Jose, or 100km by road. Its elevation ranges between 600 to 700 meters. Rara Avis borders Braulio Carillo National Park and the Zona Protectora of the La Selva Biological Station, which allows freedom of movement by the indigenous fauna.

The Waterfall Lodge at Rara Avis is a rustic lodge with eight rooms, each of which has a private bath and balcony. Each room has a double bed, single bed, and two fold-down bunks. Capacity is 16 people based on double occupancy or 32 based on quadruple occupancy. The Penal Colony Lodge (also called Albergue El Plastico) is a restored building once used as a jungle prison colony many years ago. There are now flush toilets and hot showers (shared), with kerosene refrigeration and lighting. There is a capacity of 19 in bunk beds in four rooms. A series of additional cabins in the forest, ranging from basic to luxurious, are planned for the future.

Three country-style Costa Rican meals a day are included in the basic charge. All are served family style, usually cooked over a wood stove, and include: chicken or meat, rice, black beans, tortillas, salad, fresh fruit, beverage, coffee, or hot chocolate.

There is an extensive trail system at Rara Avis, with paths that vary in difficulty as they run through primarily virgin rainforest. Groups are taken out on trails by staff biologists and naturalists, who explain different aspects of the surrounding flora and fauna.

The nearest town to Rara Avis is Las Horquetas de Sarapiqui, approximately 10 miles away. Las Horquetas can be reached from San Jose in most types of vehicles. Travel to Rara Avis from Las Horquetas is by jeep, by horseback, or by tractor-pulled cart. The trip usually takes 1-3 hours, depending on the method of transportation. The journey involves fording two rivers and traveling over 10 miles of clay, mud, and wood roads.

The costs for staying at the Rara Avis Jungle Lodges in 1989 were as follows:

Penal Colony Lodge
$35/person/night, plus transportation

Waterfall Lodge
Single room with 1 person: $75/night
Double room with 2 people: $50/person/night
Triple room with 3 people: $40/person/night
Quadruple room with 4 people: $40/person/night

Transportation from Las Horquetas to Rara Avis was quoted at $10 per person round trip, regardless of the method. Additional information and current prices can be obtained by writing to Amos Bien at the above address.

Green Turtle Research Station

Caribbean Conservation Corporation
P.O. Box 2866
Gainesville, Florida 32602 USA
Telephone: (904)-373-6441
Telex: 387530 CCC TAS

Or

Caribbean Conservation Corporation
Apartado Postal 6975-1000
San José
Costa Rica
Telephone: 38-80-69
Telex: 3015 CESTA CR

The John H. Phipps Green Turtle Research Station is located on the east (Caribbean) coast of Costa Rica, near the village of Tortuguero. Established by the Caribbean Conservation Corporation in 1954, the station has provided support facilities to scientists studying the Atlantic green sea turtle. Although previously open only during the turtle nesting season (July-

September), the GTR Station is now open continuously all year and is available to scientists, researchers, and student groups.

Several ecosystems are included in the Tortuguero area, including rainforests. Situated in the Caribbean lowlands, it is classified as very wet tropical rainforest and receives approximately 6,000mm of rainfall each year. The GTR Station is located in close proximity to Tortuguero National Park which comprises 21,000 hectares of land to its south and east, and to Barra del Colorado Wildlife Refuge which contains 92,000 hectares of land and is located to its north and west.

The following facilities are available to researchers at the Green Turtle Research Station:

- Private and dormitory rooms with capacity to house 16-18 people.
- A camping area (under construction) to accommodate groups of up to 30 people.
- A kerosene refrigerator and kerosene freezer.
- A variety of boats and motors for individual and group use.
- 110V electricity (provided by a diesel generator).
- A propane-fueled drying box that accepts standard plant presses.
- UHF radio communication (24 hr) with San Jose.
- Access to a telephone
- Full-time bilingual station manager.
- Trail guides, boat guides, and laborers available for employment at reasonable rates.
- A list of research publications resulting from studies at the GTR Station.
- Lists of flora and fauna of the Tortuguero area.

The quickest way of reaching the GTR Station from San Jose is by air. There is a landing strip in front of the station and single-engine aircraft can be chartered from the general aviation airport in the San Jose suburb of Pavas. An alternative method is to take a train or bus to the coastal port city of Limon. From Limon one must get to the river port town of Moin (approximately 5 miles distant). From Moin, there are several boats that travel to Tortuguero on a weekly basis. More detailed information including schedules and prices can be obtained from the Caribbean Conservation Corporation. The CCC also has a list of room and boat rental rates, upon request.

Tortuga Lodge

Costa Rica Expeditions
P.O. Box 6941
San Jose
Costa Rica
Telephone: 22-03-33

The Tortuga Lodge is a tourist facility located directly across from the Green Turtle Reesearch Station (see previous section) on the other side of Tortuguero Lagoon near the Caribbean coast. It specializes in fishing trips, but also serves many tourists who wish to visit the turtle nesting beaches near Tortuguero. Costa Rica Expeditions offers a number of 'package' tours to the Tortuga Lodge of various duration and means of transportation. Costa Rica Expeditions can be contacted at the above address for more details and price quotes.

Marenco Biological Station

The Marenco Biological Station is a private reserve located on the northwestern coast of the Osa Peninsula. This peninsula juts out into the Pacific Ocean in southern Costa Rica, near the border with Panama. The station contains 500 hectares of largely primary forest, and offers a diversity of tropical ecosystems. Access to Marenco is by plane or boat. Marenco's location facilitates excursions to areas such as Corcovado National Park and Isla del Caño Biological Reserve. Marenco Biological Station is used by tourists, students, and researchers. Additional information can be obtained by writing to:

Marenco Biological Station
P.O. Box 4025
San Jose 1000
Costa Rica, Central America

The San Jose office for the Marenco Biological Station is at:

Edificio Cristal 2° . Piso
Avenida 1, Calles 1 y 3
Telephone: 21-15-94
Telex: 3534 Horizo CR

San Jose

Upon arrival at Juan Santamaria Airport, immigration and customs checks were quick and orderly. All baggage came into the claim area on conveyors and was taken off by the airport workers. The Customs counters were very close to the baggage claim. The airport employees also carried the baggage from Customs to a conveyor belt which moved up one floor, parallel to a flight of stairs. At the top of the stairs, your boarding pass/ticket envelope (to which the adhesive baggage claim label is stuck) is given to another airport employee who takes the bags off the vertical conveyor and checks the claim number. The exit to the outside is immediately at hand, usually with many taxis waiting.

I used a *colectivo* or shared taxi in the form of an orange VW van. It took approximately 25 minutes to get to downtown San Jose from the airport at a cost of 550 colones (approximately $8).

San Jose has many *pensiones*, small hotels, and luxury hotels, with prices that vary accordingly. Two of the *pensiones* commonly used by biologists are the Galilea and the Johnson. I stayed at the Hotel Plaza located on the Avenida Central between Calle 2 and Calle 4, next to a store called the Palacio de los Palacios. A single room with a double bed, private bathroom with sink, toilet, and hot water shower, plus a television set cost approximately $18 per night. There was no air conditioning (or need for it) as it got cool during the evenings. There was a moderately priced restaurant on the ground floor of the hotel.

The hotel staff was friendly and helpful, and I had no problems with theft or security. The management was willing to exchange currency at the official rate. The Hotel Plaza is conveniently located across from the Banco Central de Costa Rica, and near many stores and restaurants. San Jose is a very noisy city and one should try to get a hotel room as far away from the street as possible.

Taxis in San Jose were basically inexpensive (less than $3 to most places), but one should establish the fare with the driver before getting in so as to avoid confusion later.

Like any large city, San Jose has areas dangerous for people to walk around in. Some sidewalks are narrow, so expect a fair amount of crowding and jostling. Tourists are continually approached by money changers that offer only a small difference between the official rate.

Comments On Costa Rica

Many factors combine to make Costa Rica one of the most favored places of tourists and biologists alike. The size and terrain of the country provide easy access to a variety of ecological habitats within short distances. The absence of any army and long standing stability of the political system creates a comfortable atmosphere for first time visitors to Latin America and encourages long-term research projects. The extensive amount of land held as national parks or protected areas throughout the country offers an extensive variety of locales to campers, backpackers, or birdwatchers. This, combined with the low cost of food and lodging make Costa Rica an ecologically interesting and economically favorable place to visit.

Books

THE NEW KEY TO COSTA RICA - 'ALL YOU NEED TO KNOW!'
Blake, Beatrice and Anne Becher
1987. Publications in English

PARQUES NACIONALES COSTA RICA NATIONAL PARKS
(Spanish/English)
Boza, Mario A.
1988. Editores Heliconia

TROPICAL NATURE
Forsyth, Adrian and Ken Miyata
1987. Charles Scribner's Sons

COSTA RICA
Glassman, Paul
1988. Passport Press

GUANACASTE NATIONAL PARK: TROPICAL, ECOLOGICAL AND CULTURAL RESTORATION
Janzen, Daniel H.
1986. Editorial Universidad Estatal A Distancia

COSTA RICAN NATURAL HISTORY
Janzen, Daniel H. (Ed.)
1983. University of Chicago

TODO-All COSTA RICA (Spanish/English)
Navamuel, Ricardo Vilchez

LIFE ABOVE THE JUNGLE FLOOR
Perry, Donald
1986. Simon and Schuster, Inc.

DEL AMAZONAS A LA ISLA DEL COCO
Ramirez, Miguel A.

A GUIDE TO THE BIRDS OF PANAMA, WITH COSTA RICA, NICARAGUA, AND HONDURAS
Ridgely, Robert S. and John Gwynne, Jr.
1989. Princeton University Press

THE COSTA RICA TRAVELER
Searby, Ellen
1988. Windham Press

A NATURALIST IN COSTA RICA
Skutch, Alexander F.
1971. University of Florida Press

A GUIDE TO THE BIRDS OF COSTA RICA
Stiles, Gary and Alexander Skutch
1989. Cornell University Press

COSTA RICA

There are a number of bookstores in San Jose that sell many of the titles listed above, as well as excellent maps of Costa Rica and its national parks. Attractive conservation posters are also available.

Universal (Department store with large book department)
Located between Calle Central and Calle 1, with entrances on Avenida Central and Avenida 1.

Libreria Lehmann (Large book store)
Located on Calle 3, between Avenida Central and Avenida 1.

The Bookshop (Small bookstore)
Located on Avenida 1, between Calle 1 and Calle 3.

Maps

In addition to the bookstores listed above, detailed topographic maps of specific areas of Costa Rica can be obtained from the Instituto Geografico. This is located in the building that houses the Ministerio de Obras Publicas y Transportes, near the Plaza Viquez. Once there, one should go to the office of Fotos y Mapas. Large albums contain samples of the maps that can be purchased there.

Domestic sources of maps of Latin America are listed on pages 290 and 291.

Tourist Information Sources/Travel Agencies

Instituto Costarricense de Turismo
Apartado 777
San José
Costa Rica
Telephone: 23-17-33
Telex: 2281 "INSTUR"

Cámara Nacional de Turismo
Centro Comercial, 2(o floor)
Apartado 828
2000 San José
Costa Rica
Telephone: 31-15-58

LACSA (Lineas Aereas Costarricense/Costa Rican Airlines)
Telephone: 1-800-22-LACSA(in the USA and Canada)
Costa Rica Expeditions
P.O. Box 6941
San José
Costa Rica
Telephone: 22-03-33

Conservation Organizations

Tropical Science Center
Apartado 8-3870
1000 San José, Costa Rica
Central America
Telephone: 22-62-41

Fundación Neotropica
Apartado Postal 236-1002
Paseo de los Estudiantes
San José, Costa Rica

Caribbean Conservation Corp.
P.O. Box 2866
Gainesville, FL 32605 USA
Telephone: (904)-373-6441

Asociación de Estudiantes De Biología
Universidad De Costa Rica
AA 818
San José, Costa Rica

Asociación Costarricense Para La Conservación De La Naturaleza
Apartado 8-3790 B. Amon
San José, Costa Rica

Amigos De La Naturaleza
AA 162
Guadeloupe, Costa Rica

COSTA RICA

Scientific Organizations/Institutions

Museo Nacional De Costa Rica
Calle 17, Avenida Central y 2
Apartado 749
San José 1000, Costa Rica
Telephone: 21-02-95

Universidad De Costa Rica
Ciudad Universitaria 'Rodrigo Facio'
San Pedro de Montes de Oca
San José, Costa Rica
Telephone: 25-55-55   Telex: UNICORI

    Museo de Entomología
    Facultad de Agronomía
    Universidad De Costa Rica

    Museo de Zoología
    Depto. de Biología
    Universidad De Costa Rica

Centro Agronomico Tropical de Investigación y Enseñanza (CATIE)
Turrialba, Costa Rica
Telephone: 56-64-31
Telex: 8005

Organization For Tropical Studies
P.O. Box DM
Duke Station
Durham, N.C. 27706 USA
Telephone: (919)-684-5774

    or

Organización Para Estudios Tropicales
Apartado 16
Universidad de Costa Rica
San José
Costa Rica
Telephone: 36-66-96

Monteverde Institute
c/o Council on International Educational Exchange
205 East 42nd Street
New York, N.Y. 10017

INBio (Instituto Nacional de Biodiversidad)
Dr. Rodrigo Gomez, Director
Santo Domingo de Heredia
3100 Costa Rica

CHAPTER 1G   PANAMA

General Information

Area: 77,000 sq km (29,700 sq miles)

Population: 2,370,000

Capital: Panama City (population: 439,000)

Language: Spanish

Currency: Balboa (1 Balboa = 100 centesimos)

        Coins: 1, 5, 10, 25 and 50 centesimos, 1, 5, 10, and 100 balboas (U.S. coins are also legal tender)
        Notes: 1, 2, 5, 10, 20, 50 and 100 U.S. dollars (there are no Panamanian bank notes)
        Exchange Rate: The value of the balboa is kept equivalent to that of the U.S. dollar

National
Airlines: Air Panama

Telephone   Panama 507
Codes: No city codes are needed

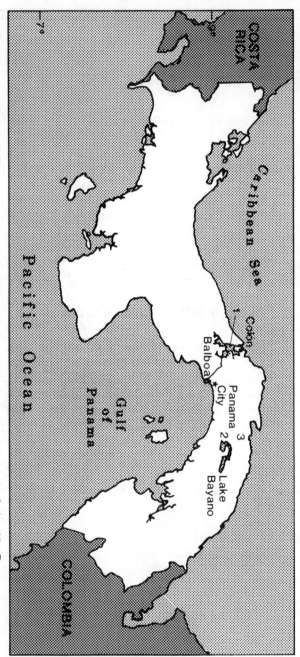

1 – Smithsonian Tropical Research Institute on Barro Colorado Island 2 – Isla Maje Scientific Reserve
3 – Nusagandi Biological Station

Introduction

The Republic of Panama is an isthmus that connects North America with South America. Panama is bordered on the east by Colombia and on the west by Costa Rica. To the north is the Carribean Sea and to the south the Pacific Ocean. The Panama Canal is located roughly in the center of the country and cuts through in approximately a northwest to southeast direction. The republic's two largest cities, Panama and Colon, are located at the southeast and northwest ends of the canal, respectively. Panama is approximately the size of South Carolina.

Panama is quite mountainous, with the Cordillera Central occupying most of the central portion of the country west of the canal. The range continues on the eastern portion along the Caribbean coast as the Serrania de San Blas and the Serrania del Darien. Tropical rainforest exists on much of the land between the coasts and the mountains due to the sparse population. At the country's western boundary with Costa Rica is located the La Amistad National Park, while near the eastern border with Colombia in the Darien Biosphere Reserve. Almost half of the population lives in the areas of Panama City and Colon. The major concentration of the rural inhabitants live in the provinces of the southwestern part of the country.

My evaluation of rainforest sites in Panama is based on information obtained and observations made during August 1988. At this time I visited the lowland rainforest of Barro Colorado Island (Smithsonian Tropical Research Institute). Information is also presented about the Isla Maje Scientific Reserve, and the Nusagandi Biologcial Station.

Barro Colorado Island

Contact

    Smithsonain Tropical Research Institute

Address

    APO Miami 34002
    Telephone: 62-31-33, 62-32-24 (Tivoli/Tupper Center)
                    62-30-49

    Or

    STRI Development Office
    Arts and Industries Bldg., Suite 2207
    900 Jefferson Drive
    Smithsonian Institution
    Washington, D.C. 20560 USA
    Telephone: (202)-786-2817

Description

    Barro Colorado Island (BCI) was formed with the creation of Gatun Lake (1910-1914), following the damming of the Chagres River. Together with adjacent peninsulas and other small islands it is part of the Barro Colorado Nature Monument under the custodianship of the Smithsonian Tropical Research Institute (STRI). BCI has been a biological reserve with laboratory facilities and lodging for scientists since 1923. Today BCI hosts hundreds of tropical biologists and students each year. Tourists and naturalists are also allowed on the island, but only on a 'space available' basis with preference given to scientists and students.

Location

    Barro Colorado Island is situated at approximately 9° North latitude. It is located near the center of Gatun Lake in the Canal Area (formerly the Canal Zone) of Panama.

Logistics

All visits to BCI have to be arranged in advance with the Visitor Services Office at the Tivoli/Tupper Center. All research must be approved before arrival. Research applications are reviewed during March, June, September, and December. Day visits can be arranged by telephone. To reach the Tivoli/Tupper Center (No. 401 Roosevelt Avenue, Ancon) you should tell the taxi driver that you want to be taken to the site of the old Tivoli Hotel on Martyr Avenue, about $1\,^1\!/_2$ blocks from the DENI station. A taxi from the Omar Torrijos Airport outside of Panama City to the Tivoli Building should take 25-35 minutes and cost $20-$25 (August 1988).

Once at the Tivoli/Tupper Center, the necessary papers can be filled out permitting visitors to enter BCI. From Ancon (where the Tivoli/Tupper Center is located) one must take a bus to Gamboa, and then a boat from Gamboa to BCI. Arrangements can usually be made through the Visitor Services Office to have extensive luggage or equipment transported by van directly to the boat in Gamboa. The boat meets two buses from Ancon daily. A schedule of arrivals and departures of buses and the ferry can be obtained from the Visitor Services Office and is posted at BCI. The bus station in Ancon is a short walk from the Tivoli/Tupper Center. BCI researchers waiting for the bus often 'kill time' in the Tropical Sciences Library next door to the Tivoli Building. The bus should take approximately 50 minutes from Ancon to Gamboa. New visitors to BCI should ask the bus driver to tell them when the bus reaches the Panama Canal Commission Dredging Division in Gamboa. This is where the BCI dock is.

The BCI dining hall, and most lodging facilities were all located near the edge of the island. There was a short walk from the ferry dock to the bottom of a hill. Most laboratories and some lodging facilities were at the top of the hill reached by stairs and an electric cart. The electric cart was on a winch system and could be summoned for use in transporting equipment. Alongside of the cart track were stairs which led up to the office and dining hall.

Forest Type

The forest at BCI is classified as Tropical Moist Forest.

The age of the forest varies from some small areas of approximately 50 years to other areas greater than 200 years.

Seasonality

The wet season at BCI lasts from approximately May through December. The dry season extends from Janaury to April. The annual rainfall is approximately 2,500-2,600 mm.

Facilities

Barro Colorado Island is forested, with the exception of several small clearings maintained at the end of certain trails, and the clearing where the laboratories and living quarters are located. Immediately at the top of the steps from the boat dock was the dining hall/dormitory. The ground floor contained the dining room, administrative office, bathrooms, and laundry area (with coin operated washers and dryers). On the second floor were rooms for 9 persons, a large shared dry-closet, and a large bathroom with sink and shower that was shared. Each room contained a bed, dresser, end table and an electric fan. Bedding, towels, and soap were also provided. Beyond the dining hall/

Dining hall (ground floor) and living quarters (second floor) at the Smithsonian Tropical Research Institute on Barro Colorado Island. (Panama)

dormitory were several other buildings capable of accommodating an additional 22 people.

Meals were served cafeteria style in the common dining room at 7:00 a.m., 12:00 p.m., and 6:30 p.m. The food was adequate and plentiful. There was usually hot water available, as well as coffee and tea. Perishable items such as milk were also available in a refrigerator in the dining hall. Bread, peanut butter, and jellies were also usually present for snacks or those who miss the regularly scheduled meals.

The living facilities had chlorinated lake water piped into them. Electricity was available both day and night in 110/220 volt single phase and 220/440 volt three phase. A telephone was also available to visitors 24 hours a day.

Researchers at BCI had access to air conditioned laboratories with sinks (untreated lake water) and electricity. The herbarium building, in addition to containing plant specimens, also held small reference collections of arthropods and vertebrates. The herbarium also contained reprints of published papers resulting from research conducted at BCI. There were also screen-houses available to researchers for maintaining animals in cages or aquariums, and for plant growth experiments. There is a fully equipped plant ecophysiology lab.

Laboratory building at the Smithsonian Tropical Research Institute on Barro Colorado Island. (Panama)

Trail System

Barro Colorado Island covers approximately 1,600 hectares. It is 4.8 km long and its highest point is 140 meters above the level of the lake. There is an excellent system of well-marked and well-mapped trails at BCI, that radiate out from the central portion and extend through all major areas of the island. More than 60 km of trails have been cut through the forest to facilitate passage and provide access to forests of different ages. Trails are clearly marked with metal tags every 100 meters.

Costs

The cost to stay on BCI and conduct research is dependent upon one's status and affiliations. Below are rates that were in effect during August 1988.

	Daily	Weekly
Researcher	$35.00	$175.00
S.I.* Staff and S.I. Post-Docs	26.00	130.00
S.I. Pre-Docs	19.00	95.00
Students/Student Assistants	17.00	85.00
Day Visitor (lunch and launch transportation)	10.00	

Non-researcher visitors are assessed fees on a case by case basis.
*S.I. = Smithsonian Institution

Anyone interested in visiting or working on BCI should write to the Tivoli/Tupper Center for the booklet *Smithsonian Tropical Research Institute - Information For Visitors*. This booklet lists facilities and staff at all the STRI laboratories in Panama, and provides historical information and guidelines for visiting and conducting research at BCI. Fee schedules, maps, bus, and boat schedules for the various STRI facilities can also be obtained at the same office. Other information of interest usually available is a current list of the visiting scientists and students at BCI and the other STRI labs.

Comments

The rainforest and its flora and fauna on BCI has probably been the most intensively studied in the Neotropics. This is not to imply however, that vast areas of biological research do not remain. The Smithsonian Tropical Research Institute provides excellent support facilities and not the least of which is the Tropical Science Library in Ancon. This, combined with quality food and lodging at reasonable rates encourages scientific researchers and facilitates the actual work. BCI has some of the most comfortable accomodations for a tropical biological station, ranking just below the La Selva Research Station (Costa Rica), in my opinion. The trail system is easy to read and follow, especially with the maps available on the island.

A knowledgeable staff of research scientists is maintained by STRI in Panama. Many of these biologists have been in residence for more than a decade and can provide great insight regarding the feasibility of various projects, as well as a wealth of information in their respective fields.

My only negative feeling regarding BCI is its location in a country having recently undergone political turmoil. As of this writing, researchers on the island itself have been relatively isolated from actions occuring in Panama City. STRI has International Mission status in Panama, giving it a neutral status similar to that of an embassy or United Nations agency. Its operations have not been greatly affected by the political situation in Panama.

Isla Maje Scientific Reserve

Contact

    Director, Gorgas Memorial Laboratory

Address

    Gorgas Memorial Laboratory
    APO Miami 34002-0012
    Telephone: 27-41-11

Description

The Isla Maje Scientific Reserve is a privately-owned island reserve that was established in 1972 as a field station for biologists. It is owned and operated by the Gorgas Memorial Laboratory, a private non-profit non-government laboratory located in Panama. Isla Maje is approximately six kilometers long by two kilometers wide, with elevations ranging from 60-100 meters. Approximately two-thirds of the 1,400 hectare reserve is primary lowland tropical forest, while less than a tenth is abandoned cattle pasture reverting back to a wild state. The remainder is a secondary scrub habitat. The facilities at the Isla Maje Scientific Reserve are available for use by both scientists and serious naturalists.

Location

Isla Maje is located in Lake Bayano, approximately 100 km northeast of Panama City. It is situated between 9°-10° North latitude.

Logistics

Isla Maje can be reached by bus or car from Panama City. It is approximately 100 km to the Bayano Bridge, which normally takes about two hours. From the bridge, a 20 minute boat ride is needed to reach the landing at Maje. Transportation can often be arranged by coordinating with the staff of the Gorgas Memorial Lab.

Forest Type

The forest at Isla Maje is transitional between Dry and Humid Lowland Tropical Forest.

Seasonality

The dry season at Isla Maje extends from approximately December through March. The wet season begins in late April, with greatest precipitation usually occuring in the month of October. The reserve receives approximately 1,500 mm of rain per year.

Facilities

The Isla Maje Scientific Reserve can comfortably accomodate 18-20 people, or slightly larger groups on a short-term basis. There is a five man field crew stationed permanently at the reserve, including someone in charge of the kitchen facilities. Researchers can arrange to have food prepared for them, or bring in their own food and do their own cooking. Sleeping accomodations are furnished by a number of rooms containing 2-3 single beds each.

Electricity is available at the field station at night.

Trail System

The 14 sq km reserve contains approximately 10 km of trails for which maps are available. There are three main trails that run in both a north-south and east-west direction, as well as in a transect that cuts through all three of the major habitat types found on the island. Access to the forest canopy is provided in the form of several 25 meter high wooden towers that have been constructed. In addition, there is an aluminum walkway that extends for approximately 60 meters which reaches a maximum height of about 15 meters.

Costs

A basic rate of approximately $10/day is charged to visitors using the station facilities. The purchase and preparation of food can be arranged for a nominal additional sum. However, the management of Isla Maje wishes to encourage the use of its facilities by researchers and will do all they can to reach a mutually agreeable working environment. All arrangements should be made prior to arrival.

Nusagandi

Contact

> PEMASKY
> Proyecto De Estudio Para El Manejo De Areas Silvestres De Kuna Yala

Address

> Apartado 2012
> Paraiso, Ancon
> Republic of Panama

Description

Nusagandi is a biological station located on a 60,000 hectare reserve called the Comarca de Kuna Yala (San Blas), which is inhabited and managed by the Kuna Indians. Although established to facilitate biological studies, naturalists and birdwatchers are allowed to use the facilities if permission is granted through PEMASKY.

Location

Nusagandi is located approximately 75km northeast of Panama City. It was constructed a short distance within the southern boundary of the Comarca de San Blas, and west of the road that extends from El Llano north to the coast. Nusagandi is situated between 9°-10° North latitude.

Logistics

To reach Nusagandi, one needs a 4-wheel drive vehicle. One drives east from Panama City on the Pan American Highway for approximately two hours until the town of El Llano. From this area there is a road that heads to the north (left), which passes near Nusagandi.

Forest Type

There is access to four different life zones at Nusagandi: Tropical Moist Forest, Tropical Wet Forest, Tropical Premontane Wet Forest, and Tropical Premontane Rainforest.

Seasonality

Annual precipitation ranges between 2,500 mm and 3,300 mm during the rainy season (from May to December), and approximately 60 mm in the dry season (from January to April).

Facilities

At Nusagandi there is a two-floor dormitory of modern construction and with beds capable of sleeping a total of 40 people. There is a kitchen with gas stoves, food storage area, and open-air dining hall capable of accomodating 30 people at a time. An arrangement can usually be made with the local people to have meals prepared. There are two sets of trails, one for public use and one for patrolling the area.

Costs

Rates charged to visitors, quoted to me in April 1988, were $10 per day. This is based on one bringing in their own food and using their own transportation. A separate charge of $8 per day should be set aside for a guide, if one is used through PEMASKY. If transportation to Nusagandi were arranged through PEMASKY, this would entail another charge to consider.

Parque Nacional Soberania
(Soberania National Park)

This national park includes the famous Pipeline Road, an area that has been the site of many STRI studies. The park is accessible from Gamboa, a small town bordering Lake Gatun. Apartments are maintained by STRI in Gamboa and may be available for researchers to rent. Permission to work in Parque

Nacional Soberania must be secured from INRENARE (Instituto Nacional de Recursos Naturales Renovables). Parque Nacional Soberania together with the Barro Colorado Nature Monument are two areas of protected rainforests that contain half of all the birds and mammals reported from Panama.

Summit Botanical Gardens

In addition to the botanical specimens, the Summit Botanical Gardens is also the site of the Center for Reproduction of Endangered Panamanian Animal Species (CEPEPE). This is a cooperative venture between INRENARE, the Panama Audubon Society, and the Mayor's Office of the Capital District of the Republic of Panama. More information on the CEPEPE project can be obtained from the Panama Audobon Society (see page 212 for address).

Panama City

I spent only one night in Panama City and this was at the Hotel Acapulco (Telephone: 25-38-32). This hotel is located on Calle 30 Este, between Avenida Cuba and Avenida Peru. The rate for a single was $22. The room was large, air-conditioned, and contained a double bed, closet, chairs, and television. There was also a private bathroom with toilet, sink, and shower with hot water. A taxi from the Hotel Acapulco to the Omar Torrijos Airport cost $15. An exit tax of $15 also had to be paid at the airport before leaving.

Comments On Panama

Panama is a beautiful country, which unfortunately was in the midst of political upheaval during my visit. As of this writing, Manuel Noriega is imprisoned in the United States. Hopefully this will have a beneficial and calming effect on the political situation in Panama. However, biologists at STRI informed me that normally scientific travel has not been affected in Panama.

Books*

PARTING THE GREEN CURTAIN: THE EVOLUTION OF TROPICAL BIOLOGY IN PANAMA (Spanish/English)
Angehr, George R.
1989. Smithsonian Tropical Research Institute

GUÍA DE LOS ÁRBOLES COMUNES DEL PARQUE NACIONAL SOBERANÍA
Angehr, George R. and Pnyllis Coley
1984. Smithsonian Tropical Research Institute

MY TROPICAL AIR CASTLE - NATURE STUDIES IN PANAMA
Chapman, Frank M.
1929. D. Appleton and Company

FLORA OF BARRO COLORADO ISLAND
Croat, Thomas B.
1978. Stanford University Press

THE BOTANY AND NATURAL HISTORY OF PANAMA
D'Arcy, William G. and Mireya Correa
1985. Missouri Botanical Garden

AVES DE LAS SELVAS PANAMEÑAS
Delgado, Francisco
1987. Fundación de Parques Nacionales y Medio Ambiente

FINDING BIRDS IN PANAMA
Edwards, Ernest P. and Horace Loftin
1971. Ernest P. Edwards (Sweet Briar, VA)

---

* Some of these titles can be purchased in the STRI Mini-Bookstore. A complete list of the books they have available can be obtained by writing to them at the STRI address provided earlier.

EL DESAROLLO DE LAS CIENCIAS NATURALES Y LA MEDICINA EN PANAMÁ
Escobar, Novencido
1987. Editorial Universitaria, Universidad de Panama

COLONIZACIÓN Y DESTRUCCIÓN DE LOS BOSQUES EN PANAMÁ
Heckadon, Stanley and A. McKay
1984. Smithsonian Tropical Research Institute

THE ECOLOGY OF A TROPICAL FOREST - SEASONAL RHYTHMS AND LONG-TERM CHANGES
Leigh, Egbert G. Jr., A. Stanley Rand and Donald M. Windsor
1982. Smithsonian Institution Press

ADAPTIVE RADIATION IN PREHISTORIC PANAMA
Ranere, Anthony and Olga Linares
1980. Harvard University Press

A GUIDE TO THE BIRDS OF PANAMA
Ridgely, Robert S.
1981. Princeton University Press

A GUIDE TO THE BIRDS OF PANAMA, WITH COSTA RICA, NICARAGUA, AND HONDURAS
Ridgely, Robert S. and John Gwynne, Jr.
1989. Princeton University Press

THE BIRDS OF THE REPUBLIC OF PANAMA
Wetmore, Alexander
1965. Smithsonian Institution

A DAY ON BARRO COLORADO ISLAND
Wong, Marina and Jorge Ventocilla
1986. Smithsonian Tropical Research Institute

UN DÍA EN LA ISLA DE BARRO COLORADO, PANAMÁ
Wong, Marina and Jorge Ventocilla
1986. Smithsonian Tropical Research Institute

Maps

The best maps of the country and an updated atlas of Panama can be purchased at the Instituto Geografico Nacional 'Tommy Guardia'. This is located across from the University of Panama on Via Simon Bolivar. Domestic sources of Latin American maps are listed on pages 290 and 291.

Tourist Information Sources/Tourist Agencies

Instituto Panameño de Turismo (IPAT)
Centro de Convenciones Atlapa
Via Cincuentenario
Apartado 4421
Panamá 5
Telephone: 26-700

Asociación Panameña de Agencias de Viajes y Turismo (APAVIT)
Apartado 2533
Panamá 3
Telephone: 26-700

Darien Compound Safaris
P.O. Box 909
Panama 9A, Panama

Conservation Organizations

Asociación Nacional para la Conservación de la Naturaleza (ANCON)
Apartado 1387
Zona 1, Panamá
Republic de Panamá

National Parks and Environmental Foundation
Apartado 278
Herrera, Panamá

Amigos De La Naturaleza
Apartado 286
David, Chiriqui, Panamá

Asociación Conservacionista De La Universidad De Panama
Apartado 6-7534
Estafeta El Dorado, Panamá

Asociación Estudantil Para La Conservación Ambiental De Panama
2797 Balboa
Panamá, Panamá

Panama Audubon Society
Box 2026
Balboa, Republic of Panama

Scientific Organizations/Institutions

Smithsonian Tropical Research Institute
APO Miami 34002
Telephone: 52-21-24 (BCI), 22-02-11 (Tivoli Building), 62-20-08 (Ancon Building), and 62-31-51 (Tropical Sciences Library)

Gorgas Memorial Laboratory of Tropical and Preventative Medicine
Avenida Justo Arosemena 35-30
Apartado 6991
Panamá 5, Panamá
Telephone: 27-41-11

Academia Nacional De Ciencias De Panamá
Apartado 4570
Panamá, Panamá

Museo De Ciencias Naturales
Avenida Cuba, Calle 29 y Calle 30
Apartado 662
Panamá
Telephone: 25-06-45

Universidad De Panamá
Ciudad Universitaria 'Dr Octavio Mendez Pereira'
El Cangrejo
Apartado Estafeta Universitaria
Panamá, Panamá
Telephone: 63-61-33

Instituto De Investigaciones Agropecuarias De Panamá (IDIAP)
Apartado 6-4391
El Dorado
Panamá, Panamá

CHAPTER 2 RAINFOREST
INFORMATION SOURCES

Introduction

The following is a selected list of references that provide information about tropical forests, primarily in the neotropical region of the world. These references include books, journals and magazines, maps, and organizations. Most of the organizations listed are conservation-oriented and based in the United States. Select Latin American organizations have been listed in Chapter 1, following the description of rainforest sites in their corresponding country.

Sources on this list are aimed at both laymen and biologists. They provide information on rainforests, the countries in which they occur, flora and fauna, social and political issues, and scientists currently conducting research in tropical biology. This list is intended to serve as a starting point for those readers interested in knowing more about rainforests.

Books are listed alphabetically according to author, and classed under several categories. Annotations briefly describing the book contents will appear below the bibliographical information in select sections. Most of the books contain information or accounts that are applicable to rainforests or the Neotropics in general. More specialized works pertaining to specific countries (such as guidebooks or books describing national parks), have been listed in the book section of that particular country in Chapter 1. Some of the books listed in this chapter refer to countries that were not covered in Chapter 1, such as Brazil and Guyana. However, others have already appeared following their appropriate country description.

Certain older books listed may be out of print and difficult to find. The following company provides a free search service to locate books, with no obligation to buy:

Out-Of-State Book Service
Box 3253
San Clemente, California 92672 USA
Telephone: (714)-492-2976

Certain companies deal primarily or solely with natural history titles. Several such companies are the following:

E.W. Classey Ltd.
Park Road, Faringdon
Oxon, SN7 7DR
England

Flora & Fauna Books
P.O. Box 15718
Gainesville, Florida 32604 USA
Telephone: (904)-373-5630

Donald E. Hahn
Natural History Books
Box 1004,
Cottonwood, Arizona 86326-1004 USA
Telephone: (602)-634-5016

Doug Kibbe
Box 34
Maryland, New York 12116 USA

Patricia Ledlie - Bookseller
1 Bean Road
P.O. Box 90
Buckfield, Maine 04220 USA
Telephone: (207)-336-2969

Melvin Marcher, Bookseller
6204 N. Vermont
Oklahoma City, Oklahoma 73112 USA

Natural History Book Service Ltd.
2 Wills Road
Totnes, Devon TQ9 5XN
Great Britain

Sandhill Crane Press
2406 N.W. 47th Terrace
Gainesville, FL 32606 USA

Selva Natural History Books
P.O. Box 5213
Tyler, TX 75712 USA

A more complete list of book publishers and booksellers dealing with natural history titles and antiquarian books can be found in *A. B. BOOKMAN'S YEARBOOK: The Specialist Book World Annual* which is put out by A. B. Bookman Publications, Inc.

The following company has a separate catalog listing books related to Latin America:

Karl Franger Books
2434 St. Lawrence Street
Vancouver, B.C.
Canada V5R 2R5

For those readers interested in information on the archaeology and anthropology of Latin America, the following companies are recommended:

Flo Silver Books
8442 Oakwood Court North
Indianapolis, Indiana 46260 USA
Telephone: (317)-255-5118

Adventures Unlimited Press
P.O. Box 22
Stelle, Illinois 60919-9989 USA
Telephone: (815)-253-6390

Companies specializing in travel guide books that also have a wide selection of maps are:

World Wide Books And Maps
736 A Granville Street
Vancouver, B.C.
Canada V6Z 1G3

Travel Books Unlimited
4931 Cordell Avenue
Bethesda, Maryland 20814 USA
Telephone: (301)-951-8533

NATURAL HISTORY AND ECOLOGICAL BOOKS FOR LAYMEN

THE TROPICS
Aubert de la Rue, Edgar, Francois Bourliere and Jean-Paul Harroy
1957. Alfred A. Knopf

This book, although outdated, still provides useful information on the world's tropical forests. It contains interesting black and white photos as well as color plates.

JUNGLES
Ayensu, Edward S. (Editor)
1980. Crown Publishers

A large format book, richly illustrated with color photographs and drawings. Discusses early exploration, flora and fauna, coevolution, commercial aspects, and the role of man. Examples are cited from tropical forests worldwide.

THE LAND AND WILDLIFE OF SOUTH AMERICA
Bates, Marston
1964. Time-Life Books

This book is part of the Life Nature Library series. Although outdated, it still provides much useful general information on South America and its fauna. Contains both older black and white and color photographs.

THE WINDWARD ROAD
Carr, Archie
1979 (First published in 1955). University Presses of Florida

A popular account of the efforts of famed scientist and naturalist Archie Carr, as he conducts his studies of sea turtles. Much of his work took place at Tortuguero (Costa Rica), where there is now a national park and the Green Turtle Research Station operated by the Caribbean Conservation Corps (see page 183).

RAINFORESTS: A GUIDE TO RESEARCH AND TOURIST FACILITIES AT SELECTED TROPICAL FOREST SITES IN CENTRAL AND SOUTH AMERICA
Castner, James L.
1990. Feline Press (P.O. Box 7219, Gainesville, FL 32605)

By describing in detail specific accessible rainforest sites from countries throughout Latin America, this unique title allows the layman to make an informed decision as to which is best to visit for his/her purposes. The characteristics of each site discussed are location, logistics, forest type, seasonality, facilities, trail system, costs, contact and address. Other features of the book include a large, partially annotated, bibliography of rainforest literature as well as descriptions of conservation organizations actively involved in tropical forest preservation. A chapter called 'Hands-On' Organizations describes alternatives available to non-scientists who wish to actually participate in the field on research projects.

IN THE RAINFOREST
Caufield, Catherine
1984. University of Chicago Press

   A well researched and informative account of the socio-economic and political problems that affect how tropical forests are currently being used. Information is presented from rainforests all over the world.

MY TROPICAL AIR CASTLE - NATURE STUDIES IN PANAMA
Chapman, Frank M.
1929. D. Appleton and Company

   Chapman describes the nature, wildlife, and forest of Barro Colorado Island in Panama during the early 1920's. He offers a clear description of the flora, fauna, and condition of the island which was formed by the damming of the Chagres River during the construction of the Panama Canal. Working there as a scientist at the Institute for Research in Tropical America, Chapman also served as the Curator of Birds for the American Museum of Natural History.

JACQUES COUSTEAU'S AMAZON JOURNEY
Cousteau, Jacques-Yves and Mose Richards
1984. Harry N. Abrams, Inc.

   This large format 'coffee table' book is filled with high quality, color photographs taken throughout the Amazon Basin. The text and photos document an expedition and voyage made by Cousteau and his crew. Using boats, rafts, trucks, an airplane, and a helicopter, the Cousteau team traveled from the headwaters of the Amazon in the Andes Mountains of Peru, down the river itself, and then along the major tributaries.

SOUTH AMERICA AND CENTRAL AMERICA - A NATURAL HISTORY
Dorst, Jean
1967. Random House

This book treats the natural history of the geographical regions of South America, with only one of 16 chapters devoted to Central America. Similar in form and content to the Life Nature Library series, it is in large format and contains vintage black and white photos.

PEOPLE OF THE TROPICAL RAIN FOREST
Denslow, Julie S. and Christine Padoch (Editors)
1988. University of California Press

This large format book with excellent color photographs takes a look at the inhabitants of tropical forests throughout the world. Over 20 reknown contributors have written chapters which concentrate on tribal peoples, small landholders and immigrants to the rainforest, business and logging in tropical forests, and government policies on land use.

EXTINCTION - THE CAUSES AND CONSEQUENCES OF THE DISAPPEARANCE OF SPECIES
Ehrlich, Paul R. and Anne H. Ehrlich
1981. Random House

A well researched and written work that discusses the current accelerated rate of extinction of species. Sections cover the importance and benefits derived from conserving species, how humans directly and indirectly endanger species, and the strategy and tactics used for conservation. One subsection specifically treats the 'politics of extinction'.

TROPICAL NATURE: LIFE AND DEATH IN THE RAINFORESTS OF CENTRAL AND SOUTH AMERICA
Forsyth, Adrian and Ken Miyata
1987. Charles Scribner's Sons

A beautifully written account of the natural history and interrelationships occurring in neotropical forests. Much of the text is derived from the first-hand experiences of the authors who worked extensively in Costa Rica and Ecuador.

ANIMALS OF SOUTH AMERICA
Gilbert, John (From the original text of Felix Rodriguez de la Fuente)
1978. Orbis Publishing Limited (London)

The majority of this large format book is devoted to rainforest fauna, although animals of the Andes, pampas, steppes, and savannahs are also mentioned. There are many color photographs which vary in quality. It provides a good introduction to both vertebrate and invertebrate neotropical fauna.

SAVING THE TROPICAL FORESTS
Gradwohl, Judith and Russell Greenberg
1988. Island Press

Beginning with an introduction that discusses different aspects of deforestation, the authors go on to present actual case studies involving tropical forests. These case studies are seldom more than several pages long and, although examples are provided from all over the world, major emphasis is on Latin American forests. The section presenting the case studies is divided into four chapters covering: Forest Rreserves, Sustainable Agriculture, Natural Forest Management, and Tropical Forest Restoration.

LESSONS OF THE RAINFOREST
Head, Suzanne and Robert Heinzman (Editors)
1990. Sierra Club

This book combines works from two dozen different contributors and presents them under five broad headings: The Tropics, Bioregional History, Bad Management, Wise Management, and Saving The Forest And Ourselves. Individual chapters under these headings discuss a wide range of topics, citing examples from tropical forests around the world. Contributors represent scientists, conservationists, and others who have a long history of association with rainforests.

A MANUAL FOR BIRD WATCHING IN THE AMERICAS
Heintzelman, Donald S.
1979. Universe Books

Although 10 years old, this work provides much useful information about bird watching in general. Two separate chapters deal with birds and birding locations in Central and South America, respectively. There are a number of black and white photos and some color photos and some color plates provided, although the book is not a photo-identification guide.

GREEN INHERITANCE - THE WORLD WILDLIFE FUND BOOK OF PLANTS
Huxley, Anthony
1985. Anchor Press/Doubleday

This large format book, put out as a conservation message by the World Wildlife Fund, combines beautiful color photography with equally impressive color drawings of the plant world. The author discusses the importance of plants to the environment, their cultivation and use as dietary staples, their importance to the world as sources of medicine, and the various other functions they serve including ornamental uses.

THE EMERALD REALM
Krassman, Steven
(To be published in 1990). National Geographic Society

A NEOTROPICAL COMPANION - AN INTRODUCTION TO THE ANIMALS, PLANTS, AND ECOSYSTEMS OF THE NEW WORLD TROPICS
Kricher, John C.
1989. Princeton University Press

One of the most thorough, well written works that introduces and explains rainforest ecology, flora, and fauna. Equally appropriate for serious layman or biologist, Kricher covers the abiotic and biotic factors essential to understand the functioning and interactions occurring within a tropical forest. Points are verbally illustrated citing examples and experiences from specific New World forests. There is an excellent glossary and reference section.

TREES OF LIFE: SAVING FORESTS AND THEIR BIOLOGICAL WEALTH
Miller, Kenton and Laura Tangley
1990. World Resources Institute

This book is the second in the WRI's Guides To The Environment Series, which attempts to clearly explain complex environmental issues. *TREES OF LIFE* starts with a general overview and history of the problem of deforestation. The current situation is discussed next, using Amazonia as an example. Deforestation in other geographical areas including the U.S. is covered, as well as successful re-forestation programs. Governments' roles in stopping deforestation is discussed, as are avenues available to the reader as just one person wishing to act.

THE ENCHANTED CANOPY - A JOURNEY OF DISCOVERY TO THE LAST UNEXPLORED FRONTIER, THE ROOF OF THE WORLD'S RAINFORESTS
Mitchell, Andrew W.
1986. Macmillan Publishing Company

This book discusses one of the least known and most inaccessible aspects of tropical forests - the canopy. Flora, fauna, and people that live in close association with the upper levels of the rainforest are treated. Information is drawn from researchers and regions all over the world, supplemented with superb color photographs.

MARGARET MEE IN SEARCH OF FLOWERS OF THE AMAZON FORESTS
Morrison, Tony (Editor)
1989. Nonesuch

This large format book features the artwork of Margaret Mee, who traveled extensively throughout the Amazon region for 30 years beginning in 1956. Color reproductions of her paintings of exotic plant and wildlife, as well as photographs of the Amazon are provided. The text is based on Margaret Mee's diaries of her travels.

CENTRAL AMERICAN JUNGLES
Moser, Don
1975. Time-Life Books

This book fits the typical format of the Time-Life series. An interesting text deals with the forests, flora, and fauna found in different geographical areas of Central America. It is illustrated with many color photographs by Co Rentmeester, which are slightly outdated by today's standards, but still enjoyable to view.

THE SINKING ARK: A NEW LOOK AT THE PROBLEM OF DISAPPEARING SPECIES
Myers, Norman
1979. Pergamon Press

This work was produced as part of a project conducted by the Natural Resources Defense Council (Washington, D.C.). Myers divides the text into three major parts: 1) The problem of disappearing species, 2) Tropical moist forests, and 3) A comprehensive strategy for conservation of species. This book is filled with statistical information and draws on examples from Latin America, Southeast Asia, and Africa.

A WEALTH OF WILD SPECIES: STOREHOUSE FOR HUMAN WELFARE
Myers, Norman
1983. Westview Press

A review is presented of the contributions that wild species have made to our lives, using examples in agriculture, medicine, and industry, as well as providing a section on genetic engineering. Although written in lay form, citations are given as in scientific journals making it easy to trace back to the original source of information. The economic value and impact of examples listed provide pragmatic arguments for conservation.

TROPICAL RAINFORESTS - ENDANGERED ENVIRONMENT
Nations, James D.
1988. Franklin Watts

The author, who studies the human use of tropical ecosystems, presents an overall picture of tropical rainforests and the problems that are threatening them. Included in his discussions are some of the wonders of the rainforest, their importance to man, and discussions of the indigenous people that inhabit tropical forests.

LIFE ABOVE THE JUNGLE FLOOR
Perry, Donald
1986. Simon and Schuster

An account of the rainforest community based on first hand observations of a biologist who has worked in the forest canopy. An ingenious system of ropes and pulleys has allowed the author to study and record little known interactions of the plants and animals in the treetops. Excellent color photographs are provided. This work was conducted at Rara Avis (see page 182) near Finca La Selva in Costa Rica.

THE WORLD OF THE JAGUAR
Perry, Richard
1970. Taplinger Publishing Company

Perry describes the life and habits of the jaguar, with many quotations from different sources and researchers. The jaguar's interaction with other wild animals and with man is discussed. There are some older black and white photos of jaguars and other New World animals.

LIFE IN FOREST AND JUNGLE
Perry, Richard
1976. Taplinger Publishing Company

This book is Volume 4 in The Many Worlds of Wildlife Series. Perry describes various animals and their interrelationships from tropical forests in Latin America, Africa, Asia, and Australia, as well as from temperate forests. About one-fifth of the text is devoted to American rainforests and the major animals treated in this section are termites and army ants.

JAGUAR
Rabinowitz, Alan
1986. Arbor House

An account of the author's field experiences studying jaguars in the Cockscomb Basin area of Belize, and his successful efforts to establish the world's first jaguar preserve.

THE LIFE OF THE JUNGLE
Richards, Paul W.
1970. McGraw-Hill, Inc.

This work written for the layman is by the same author who wrote *THE TROPICAL RAIN FOREST*, which was the classic text on the subject for decades. Heavily illustrated with somewhat dated color photographs, this book provides an introduction to the animals, plants, and humans that live in rainforests worldwide.

RAIN FORESTS AND CLOUD FORESTS
Sandved, Kjell B. and Michael Emsley
1979. Harry N. Abrams, Inc.

This is a large format 'coffee table' book that discusses the flora and fauna of rainforests, and their interrelationships. It is filled with large color photographs taken in tropical forests throughout the world.

IVAN SANDERSON'S BOOK OF GREAT JUNGLES
Sanderson, Ivan T. and David Loth
1965. Julian Messner

This book is based on the experiences and observations made by Sanderson through decades of travel through the jungles of the world. Plants and animals are described, as well as indigenous peoples and explorers. It is illustrated with black and white photos and line drawings.

EXPLORING THE AMAZON
Schreider, Helen and Frank Schreider
1970. National Geographic Society

This volume recounts the authors' journey by boat from the headwaters of the Amazon River in the Andes of southeastern Peru, to the mouth of the Amazon in Brazil at the Atlantic Ocean. Produced in typical National Geographic style, the book is filled with many excellent (although somewhat dated) photos.

THE WORLD OF THE RAIN FOREST
Silverberg, Robert
1967. Meredith Press

In layman's terms, the author discusses rainforests and selected examples of various plants and animals. There are few photographs, all black and white and not of high quality.

A NATURALIST ON A TROPICAL FARM
Skutch, Alexander F.
1980. University of California Press

Skutch writes about his life and observations made from his homestead in Costa Rica. Chapters cover seasonality in the tropics and a variety of specific birds, plants, insects, and animals.

NATURE THROUGH TROPICAL WINDOWS
Skutch, Alexander F.
1983. University of California Press

This book discusses primarily the behavior of birds. The text is based on observations made by the author from and near residences he lived at in Panama, Costa Rica, Honduras, and Guatemala. Skutch's original observations provide great insight to avian behavior.

A NATURALIST AMID TROPICAL SPLENDOR
Skutch, Alexander F.
1987. University of Iowa Press

Skutch describes the beauty of the tropics based on his personal observations of several decades. He discusses various flora and fauna, concentrating heavily on the behavior of birds.

THE AMAZON
Sterling, Tom
1973. Time-Life International (Netherlands) B.U.

This is part of the Time-Life series on the World's Wild Places. It is typical of Time-Life books, produced in large format and filled with color photographs (which appear coarse by today's high standards). It provides an easily readable account of the flora, fauna, and some of the human inhabitants of the Amazon Basin.

DREAMS OF AMAZONIA
Stone, Roger D.
1985. Viking Penguin Inc.

Stone discusses the Amazon region of Brazil. His topics include the geological history of the area, early explorers and naturalists, the rubber boom and development, and rainforest research projects and the scientists conducting them. He finishes with a discussion of deforestation problems.

WHERE HAVE ALL THE BIRDS GONE?
Terborgh, John
(To be published in early 1990). Princeton University Press

This book discusses the decline of bird populations in the United States and its causes, among them tropical deforestation in Central and South America.

THE EARTH CARE ANNUAL 1990
Wild, Russell (Editor)
1990. The National Wildlife Federation and Rodale Press

This book is a collection of environmental reports from the world's leading newspapers and magazines. It covers important ecological topics including rainforests, wetlands, ozone, pesticides and others.

A DAY ON BARRO COLORADO ISLAND
Wong, Marina and Jorge Ventocilla
1986. Smithsonian Tropical Research Institute

The authors of this short book used their considerable experience gained on Barro Colorado Island (BCI) to produce this lay-oriented description of its flora, fauna, and abiotic characteristics. A major part of this book is devoted to a self-guided tour of a nature trail that was established on BCI. Both common names and Latin names are used throughout the book which is illustrated with line drawings.

ADVENTURE, TRAVEL, AND EXPLORATION BOOKS (1950 - Present)

WHERE WINTER NEVER COMES - A STUDY OF MAN AND NATURE IN THE TROPICS
Bates, Marston
1963. Charles Scribner's Sons

In this well known classic, Bates discusses a number of topics pertaining to the tropics. Man, culture, climate, diseases, and food are all covered in separate chapters as well as tropical nature and the rainforest. Discussions include examples from tropical areas throughout the world.

HIGH JUNGLES AND LOW
Carr, Archie
1953. University of Florida Press

Based on the experiences of the author during the years he was a member of the faculty at the Escuela Agricola Panamericana (Panamerican School of Agriculture) in Honduras. In his enchanting style, Carr describes both the land and people he came to know in Honduras and Nicaragua.

COUPS AND COCAINE - TWO JOURNEYS IN SOUTH AMERICA
Daniels, Anthony
1987. The Overlook Press

Descriptions and impressions are presented of the land, culture, and people of South America. The text is based on the author's visits to Peru, Bolivia, Ecuador, Brazil, Paraguay, and Chile.

THE EXPLORATION OF SOUTH AMERICA: AN ANNOTATED BIBLIOGRAPHY
Goodman, Edward J.
1983. Garland Publishing, Inc.

This work lists numerous citations on the exploration of South America, beginning with the fifteenth century and continuing through the 20th. Many are categorized to a particular geographic region. The majority of listings are from hispanic books, magazines, and journals.

AMAZON
Kelly, Brian and Mark London
1983. Harcourt Brace Jovanovich, Publishers

A descriptive piece based almost entirely in Brazil. It recounts the travels made by the authors during two several month long trips into Amazonia.

WIZARD OF THE UPPER AMAZON - THE STORY OF MANUEL CORDOVA-RIOS
Lamb, F. Bruce
1974. North Atlantic Books

True tale of Manuel Cordova and his early life among the Huni Kui Indians. It describes how Cordova learned about the rainforest and came to posess a tremendous knowledge of medicinal plants. The story is told by the author, who was guided by Cordova while working as a timber specialist in northeast Peru.

RIO TIGRE AND BEYOND: THE AMAZON JUNGLE MEDICINE OF MANUEL CORDOVA
Lamb, F. Bruce
1985. North Atlantic Books

The sequel to *WIZARD OF THE UPPER AMAZON*, Lamb continues the life story of Manuel Cordova after his return from seven years of captivity with the Huni Kui Indians. In anecdotal form, stories and remembrances of Cordova are recounted, beginning with a summary of his captivity.

THE CLOUD FOREST - A CHRONICLE OF THE SOUTH AMERICAN WILDERNESS
Mathiessen, Peter
1961. Viking Press

Recounts the author's journey of approximately seven months by various means of transportation throughout South America. The title is somewhat misleading as Mathiessen starts his trip by sailing from Brooklyn to the port of Iquitos (Peru), and then traverses the continent numerous times over varied terrain.

THE PANAMA HAT TRAIL: A JOURNEY FROM SOUTH AMERICA
Miller, Tom
1986. William Morrow and Company

This book provides a thorough and entertaining look at the Panama hat (*sombrero de paja toquilla*) trade, from its humble beginnings in the Andes of Ecuador to the final destination of the hats in the United States, Europe, and elsewhere. Miller interviews numerous people involved in all aspects of the business, from the straw growers and weavers, to the middlemen, exporters, and foreign buyers.

THE DISCOVERY OF SOUTH AMERICA
Parry, J.H.
1979. Taplinger Publishing Company

This book traces European discovery and exploration of South and Central America, Mexico, and the Antillean islands. The book is filled with excerpts from narratives often written by the explorers themselves, their contemporaries, or the indigenous people they discovered. Many black and white photographs of ancient artwork and illustrations are presented.

THE AMAZON
Schulthess, Emil
1962. Simon and Schuster

This is a large format, 'coffee table' type book featuring mostly color photographs from throughout Amazonia. Photos illustrate the geography and terrain, flora and fauna, and people and cultures. A short descriptive text is also presented.

THE RIVERS AMAZON
Shoumatoff, Alex
1978. Sierra Club Books

A travelogue of the author's journey in South America. It includes visits to Indian groups in Brazil and the Peruvian Andes.

DISCOVERING MAN'S PAST IN THE AMERICAS
Stuart, George E. and Gene S. Stuart
1969. National Geographic Society

This book discusses man's development in the Western Hemisphere, treating various cultures that have existed from the Ice Age to present. It covers the major archaeological sites in North, Central, and South America, along with information on the races that created them. Many photographs of historical cities and artifacts are provided.

THE OLD PATAGONIAN EXPRESS - BY TRAIN THROUGH THE AMERICAS
Theroux, Paul
1979. Houghton Mifflin Company

Theroux describes his trip by rail from Medford, Massachusetts, to Patagonia in southern Argentina. Interspersed among descriptions of the lands he visits are conversations with the people he meets, and his own entertaining impressions.

ADVENTURE, TRAVEL, AND EXPLORATION BOOKS (Pre-1950)

THE NATURALIST ON THE RIVERS AMAZON
Bates, Henry W.
1910. John Murray

As the remaining portion of the title states, this book is " a record of adventures, habits of animals, sketches of Brazilian and Indian life, and aspects of nature under the equator, during eleven years of travel". It recounts the travels and observations of reknowned naturalist Henry Walter Bates. This book is illustrated with excellent pen and ink or line drawings.

WALLACE AND BATES IN THE TROPICS - AN INTRODUCTION TO THE THEORY OF NATURAL SELECTION
Beddall, Barbara G. (Editor)
1969. The Macmillan Company.

This book is based on the writings of Alfred Russel Wallace and Henry Walter Bates, two naturalists who traveled throughout the tropics as commercial collectors to support themselves. Contemporaries of Charles Darwin, both men contributed greatly to the theory of evolution through natural selection. The editor has chosen selections from both authors that convey the tremendous diversity of plant and animal life they observed, and which helped them to found their theories.

EDGE OF THE JUNGLE
Beebe, William
1950 (First published in 1921). Duell, Sloan and Pearce

This is a series of essays about the flora and fauna inhabiting the tropical forest near the New York Zoological Society's Tropical Research Station in Kartabo, British Guiana.

HIGH JUNGLE
Beebe, William
1949. Duell, Sloan and Pearce

Discusses the scientific expeditions led by the New York Zoological Society to the Venezuelan Andes in 1945, 1946, and 1948. Discussions focus on the habitats and wildlife found at Rancho Grande, a biological station located in a cloud forest that forms part of Henri Pittier National Park (see page 112).

THE NATURALIST IN NICARAGUA
Belt, Thomas
1985 (Originally published in 1874). University of Chicago Press

This book chronicles the observations made by the author while employed as a mining engineer in Nicaragua from 1868-1872. His style of writing shows the attention to detail expected of a scientist, while also clearly demonstrating the author's love and enthusiasm for the field of natural history. Each chapter combines a variety of descriptions, not only of flora and fauna and their interactions, but also of geology, anthropology, and travel. Belt proposes many theories to account for his observations, which have since been verified and/or are under discussion by tropical biologists today.

ACROSS SOUTH AMERICA
Bingham, Hiram
1911 (Republished in 1976). Houghton Mifflin Company

World reknown archaeologist and excavator of Machu Picchu describes his exploration of the historic trade route between Lima, Potosi, and Buenos Aires during 1908-1909. Utilizing steamer, rail, and mule, Bingham recounts his trip through Brazil, Argentina, Bolivia, Peru, and Chile. Older copies of this book are filled with vintage black and white photographs.

LOST CITY OF THE INCAS - THE STORY OF MACHU PICCHU AND ITS BUILDERS
Bingham, Hiram
1948. Duell, Sloan and Pearce

Bingham describes all that was known at the time about the archaeological ruins of Machu Picchu and the Inca civilization that existed at the time of their construction. He relates the search for ancient Incan cities, his discovery of Machu Picchu, and the exploration of the site. Vintage black and white photographs of the ruins and their artifacts are included.

LANDS OF THE ANDES AND THE DESERT
Carpenter, Frank G.
1927. Doubleday, Page and Company

This book is part of the Carpenter's World Travels series. It relates the author's journey through Colombia, Ecuador, Peru, and Bolivia. The book is supplemented with many fine black and white photographs.

THE GREAT NATURALISTS EXPLORE SOUTH AMERICA
Cutright, Paul R.
1940. Macmillan Company

This book is divided into two sections. The first provides descriptions of the accounts of the famous early naturalists who explored South America. The second part describes a variety of the animals found in South America, including insects, fish, birds, mammals, and reptiles.

THE VOYAGE OF THE BEAGLE
Darwin, Charles R.
1988 (originally published in 1839). NAL Penguin, Inc.

This book is essentially a field journal recorded by a youthful Charles Darwin as he served as the naturalist on the H.M.S. Beagle from 1831-1836. The ship surveyed the coast of South America, providing Darwin with the opportunity to observe living and fossil flora and fauna and geological formations. Many of his observations were later used to support theories and hypotheses presented in his *ORIGIN OF SPECIES*.

THE LOST WORLD
Doyle, Arthur C.
1912. Hodder and Stoughton

Fictional account of a scientific expedition that travels through the wilderness of Brazil's jungle to reach a great rock plateau (now called a *tepui*), where prehistoric life forms can still be found.

BRAZILIAN ADVENTURE
Fleming, Peter
1960. Charles Scribner's Sons

Fleming describes his 3,000 mile journey through the Matto Grosso region of Brazil as a member of a British expedition, searching for Colonel P.H. Fawcett. His accurate account of this 1932 expedition was originally published in 1933.

THE EXPLORATION OF SOUTH AMERICA: AN ANNOTATED BIBLIOGRAPHY
Goodman, Edward J.
1983. Garland Publishers

This work lists numerous citations on the exploration of South America, beginning with the fifteenth century and continuing through the twentieth. Many are categorized to a particular geographic region. The majority of listings are from Hispanic books, magazines, and journals.

A NATURALIST IN BRAZIL
Guenther, Konrad
1931. Houghton Mifflin Company

The subtitle of this work explains that it is "The record of a year's observation of her flora, of her fauna, and her people". Translated from German, the author also supplies his own photos and drawings.

PERSONAL NARRATIVE OF TRAVELS TO THE EQUINOCTIAL REGIONS OF AMERICA DURING THE YEARS 1799-1804.
Humboldt, Alexander von
1969 (Originally published in 1851). Ayer Company Publishers

This book is a journal made by the author and famous explorer whose fields of accomplishments included navigation, botany, zoology, and geology. Among the New World locations visited by Humboldt were Venezuela, Ecuador, Peru, Cuba, and Mexico. The author recounts in meticulous detail, observations in all of the above mentioned disciplines and more.

WHITE WATERS AND BLACK
MacCreagh, Gordon
1926 (Republished in 1985). Grosset & Dunlap Publishers

The entertaining account of a scientific expedition from the point of view of a non-scientist (the author) who accompanied it. MacCreagh describes the trials and tribulations of himself and seven scientists as they spend two years in the jungle wildernesses of Bolivia, Brazil, and Colombia. Early on in the narrative the author promises "not to encumber it (the narrative) with a single item of scientific value". Some interesting black and white photos throughout the text.

JOURNEY WITHOUT RETURN
Maufrais, Raymond
1953. William Kimber

This book was compiled from notes made by the author in the six months before he disappeared in the jungles of French Guiana. It describes his adventures while traveling through French Guiana in an attempt to reach the Tumuc Humac Mountains near the Brazilian border.

THE DISCOVERY OF THE AMAZON
Medina, Jose Toribio (Editor)
1988. (Originally published in 1934). Dover Publications, Inc.

This work is a collection of papers in book form that was originally compiled by Medina in 1894. It describes the voyages of Francisco de Orellana, and his discovery and exploration of the Amazon River. A large portion of the book is devoted to the account of Orellana's voyage provided by the chronicler, Friar Gaspar de Carvajal, who accompanied him.

IN THE WILDS OF SOUTH AMERICA
Miller, Leo E.
Charles Scribner's Sons

Naturalist and field biologist Miller describes his explorations throughout South America as a member of six American Museum of Natural History expeditions, conducted from 1911 to 1916. Excellent vintage black and white photographs illustrate the book.

UP THE ORINOCO AND DOWN THE MAGDALENA
Mozans, H.J.
1910. D. Appleton and Company

This is an account of the author's journey to Venezuela and Colombia during the early 1900's. It contains a number of excellent very old black and white plates that document various portions of the exploration.

THROUGH THE BRAZILIAN WILDERNESS
Roosevelt, Theodore
1969 (First published in 1914). Greenwood Press, Publishers

Tale of Teddy Roosevelt's exploits during 1913 as he traveled with Colonel Rondon and scientists from the American Museum of Natural History through the 'hinterland' of Brazil. Written in jounal-like form by the world reknown hunter and explorer, it also offers insight into the personality of this important and colorful character from history. The book is replete with black and white photographs taken during the expedition.

NOTES OF A BOTANIST ON THE AMAZON & ANDES
Spruce, Richard
1908. Macmillan and Company

This two volume work was edited and condensed by Alfred Russel Wallace after Spruce's death. The book is a condensation of notes and letters written by Spruce during his botanical exploits from 1849 to 1864. Included in the text are old photographs as well as pencil sketches and line drawings made by the author.

INCIDENTS OF TRAVEL IN CENTRAL AMERICA, CHIAPAS & YUCUTAN
Stephens, John L.
1988 (First published in 1841). Century

This book relates the travels and discoveries of the author while acting as a special envoy on a 3,000 mile journey through the primitive areas of Mexico, Guatemala, British Honduras, Nicaragua, and Costa Rica. Beginning in 1839, Stephens and his colleagues visited the ancient Mayan cities of Copan and Palenque, among others. The book contains many fine reproductions of sketches made by Frederick Catherwood, an artist and illustrator who accompanied Stephens.

OFF WITH THEIR HEADS
Von Hagen, Victor W.
1937. Macmillan Company

This work is based on observations made by the author while staying for eight months with the head-hunting Jivaro Indians of Ecuador. It describes some of the cultural practices and social customs of what was once considered the fiercest tribe of Indians indigenous to South America.

SOUTH AMERICA - THE GREEN WORLD OF THE NATURALISTS
Von Hagen, Victor W.
1951. Eyre and Spottiswoode

This book is a collection of excerpts from 25 different authors, spanning the 16th through 20th centuries. A variety of subjects on South American natural history is discussed in selections from the most famous explorers and naturalists of the past 500 years.

Excellent copies of old line drawings are included, along with some older black and white photos.

TROPICAL NATURE, AND OTHER ESSAYS
Wallace, Alfred R.
1975 (Reprinted from an 1878 edition). AMS Press Inc.

This work was originally published in 1878, which is reflected in the style and content. Wallace discusses flora and fauna from tropical areas throughout the world, often quoting or using excerpts from other authors. The style is essay-like, and provides a glimpse of another era.

A NARRATIVE OF TRAVELS ON THE AMAZON AND RIO NEGRO
Wallace, Alfred R.
1889. Ward, Lock and Company

This work deals with Wallace's scientific collecting trip into the Amazon region during 1848-1852. This trip began in the company of Henry Walter Bates. As mentioned in the full title of the book, Wallace also gives an account of native tribes and makes observations on the climate, geology, and natural history of the Amazon Valley.

WANDERINGS IN SOUTH AMERICA
Waterton, Charles
1973 (First published in 1825). Oxford University Press

This book is concerned primarily with four journeys made by the eccentric author, from approximately 1812 to 1825. It contains some excellent reproductions of pen and ink scenes of native life and animal specimens.

TRAVELS AMONGST THE GREAT ANDES OF THE EQUATOR
Whymper, Edward
1892. Charles Scribner's Sons

Whymper discusses the exploits of himself and two companions as they travel among the mountains of Ecuador. The purpose of the journey was to research and quantify the effects of mountain sickness, and to collect botanical and zoological specimens. The book contains many excellent line drawings.

BOOKS FOR BIOLOGISTS

TROPICAL RAINFORESTS - DIVERSITY AND CONSERVATION
Almeda, Frank and Catherine M. Pringle (Editors)
1988. California Academy of Sciences

This book is the result of a symposium on 'Diversity and Conservation of Tropical Rainforests' given at the California Academy of Sciences in September 1985. Fully half of this volume reports research and conservation efforts conducted in Costa Rica, with emphasis on the Organization For Tropical Studies, La Selva Biological Station and Braulio Carillo National Park, and the Costa Rican National Parks system. The remaining half deals with primate conservation, applied aspects of tropical biology, and the value of tropical diversity and species.

PARTING THE GREEN CURTAIN: THE EVOLUTION OF TROPICAL BIOLOGY IN PANAMA
Angehr, George R.
1989. Smithsonian Tropical Research Institute

This large, finely illustrated booklet contains text in both English and Spanish and was published in conjunction with an exhibition of the same name. This exhibition celebrated the opening of the Tupper Research and Conference Center of the Smithsonian Tropical Research Institute (STRI), Panama City, Panama, August 1989. This booklet describes the development of tropical biology and the establishment of the STRI facilities on Barro Colorado Island and in Panama. Past and current STRI research projects are discussed. An excellent bibliography of nearly 150 published research articles resulting from work done at STRI facilities is included.

TROPICAL WILDLIFE IN BRITISH GUIANA
Beebe, William, G. Inness Hartley and Paul G. Howes
1917. New York Zoological Society

This book describes the general aspects of the rainforest and the animal life in the Bartica District of British Guiana. It also presents the results of studies by the scientific staff of the New York Zoological Society stationed at the Tropical Research Station in Bartica during 1916. These results are in the fields of entomology, ornithology, and embryology. There are many black and white photos and some line drawings.

A DIRECTORY OF NATURAL RESOURCE MANAGEMENT ORGANIZATIONS IN LATIN AMERICA AND THE CARIBBEAN
Buckley-Ess, Julie (Editor)
1988. Partners of the Americas

This book has collected the pertinent information (contact, address, telephone, telex, and description) of the major organizations involved with natural resources and conservation in the Neotropics. Organized by country, data are further grouped by government organization, non-government organization, and educational organizations. This book is a necessity for anyone planning conservation work in Latin America and can be of enormous aid to scientists and researchers. (Note: This book is available through Partners of the Americas, see page 302.)

RAINFORESTS: A GUIDE TO RESEARCH AND TOURIST FACILITIES AT SELECTED TROPICAL FOREST SITES IN CENTRAL AND SOUTH AMERICA
Castner, James L.
1990. Feline Press (P.O. Box 7219, Gainesville, FL 32605)

This unique work methodically describes and evaluates neotropical forest sites in seven Latin American countries with facilities capable of serving as bases for field research. Among the features of each site discussed are location, logistics, forest type, seasonality, facilities, trail system, costs, contact and address. This book also contains a large, partially annotated, bibliography of rainforest related works, as well as a chapter on sources of funding for biologists. Many U.S. and Latin American conservation organizations are mentioned.

TROPICAL RAIN-FOREST: THE LEEDS SYMPOSIUM
Chadwick, A.C. and Stephen L. Sutton (Editors)
1984. Leeds Philosophical and Literary Society

This book is composed of 22 manuscripts, based on posters that were presented at the symposium titled The Tropical Rain Forest: Ecology and Resource Management, held at Leeds University in 1982. The volume is broadly divided into sections treating Community Structure, Nutrient Cycling, Maintenance of Diversity, and Resource Management.

THE LIFE OF PLANTS
Corner, E.J.H.
1964. The World Publishing Co.

Corner describes the evolution, biology, and natural history of plants in this work. Many of the examples used to illustrate themes in the text are of tropical origin.

VERTEBRATE ECOLOGY IN THE NORTHERN NEOTROPICS
Eisenberg, John F. (Editor)
1979. Smithsonian Institution Press

This volume brings together the work conducted in Venezuela by a number of investigators associated with the Smithsonian Institution. Twenty contributors have provided 17 articles of mammalian research presented in the following sections: Habitats And Distribution Patterns; The Edentata And Marsupialia; The Primates; Bats, Carnivores, And Rodents; A Comparison Of Llanos And Rainforest Mammal Faunas; Avian Studies; and Reptile Studies.

POPULATION BIOLOGY
Emmel, Thomas C.
1976. Harper and Row, Publishers

An excellent college level textbook on population biology. One of the chapters provides a useful discussion and comparison of seasonality and populations in tropical and temperate climates and zones. Other sections, such as predator-prey interactions, rely heavily on examples from the tropics.

ELEMENTS OF TROPICAL ECOLOGY - WITH REFERENCES TO THE AFRICAN, ASIAN, PACIFIC AND NEW WORLD TROPICS
Ewusie, J. Yanney
1980. Heinemann Educational Books Ltd.

This work was written as an introductory text to tropical ecology. Chapters are divided under the two general headings of Broad Ecological Tropical Features And Methods Of Study and Tropical Ecosystem Types And Problems Of Conservation. Although the subtitle implies that references are made to the tropics throughout the world, the majority of examples used in the text appear to come from West Africa.

FRAGILE ECOSYSTEMS - EVALUATION OF RESEARCH AND APPLICATIONS IN THE NEOTROPICS
Farnworth, Edward G. and Frank B. Golley (Editors)
1974. Springer-Verlag

This book is a report resulting from The Institute of Ecology (TIE) workshop on tropical ecology. Its broad purpose is to determine research approaches that will enhance our understanding of tropical ecosystems. The workshop participants were grouped into the following teams: Tropical Population Ecology, Tropical Ecosystem Structure and Function, Recovery of Tropical Ecosystems, Ecological/Technological Interactions, Impacts of Regional Changes on Climates and Oceans, and Mechanisms to Support and Encourage Research and Education in Tropical Ecology.

BIOGEOGRAPHY AND ECOLOGY IN SOUTH AMERICA
Fittkau, E.J., J. Illies, H. Klinge, G.H. Schwabe, and H. Sioli (Editors)
1969. Dr. W. Junk N.V. Publishers The Hague

This rather unusual two volume set presents various aspects of South American biogeography and ecology. However, the different sections of the book are written in one of the following four languages: German, English, Portuguese, Spanish. Text in English and German predominates. Each article is followed by a summary that appears in three languauges: English, German, and Portuguese or Spanish.

THE EQUATORIAL RAIN FOREST: A GEOLOGICAL HISTORY
Flenley, John R.
1979. Butterworth and Company Ltd.

In this work, the vegetation history of equatorial Africa, Latin America, and Indo-Malesia is discussed, based primarily on geological history and pollen evidence. One section treats the quaternary vegetation of Latin America in depth, while others discuss seral changes, the influence of man, and present trends and prospects.

COEVOLUTION
Futuyma, Douglas J. and Montgomery Slatkin (Editors)
1983. Sinauer Associates Inc.

This book utilizes 23 contributors (several of which are noted tropical biologists) to discuss the various aspects of coevolution. A sampling of chapters includes Coevolution and Mimicry, Coevolution and Pollination, Parasite-Host Coevolution, Evolutionary Interactions Among Herbivorous Insects and Plants, the Dispersal of Seeds by Vertebrates, and others. Many examples are given discussing neotropical species.

FOUR NEOTROPICAL RAINFORESTS
Gentry, Alwyn H. (Editor)
(To be published in fall 1990). Yale University Press

According to Yale University Press, this book will bring together information on four of the most studied tropical forest sites in the Neotropics: La Selva (Costa Rica), Barro Colorado Island (Panama), a site near Manaus (Brazil), and Manu (Peru).

Many different scientists will contribute articles which will deal with both plant and animal communities.

COEVOLUTION OF ANIMALS AND PLANTS
Gilbert, Lawrence E. and Peter H. Raven
1975. University of Texas Press

This work contains a collection of papers that were originally presented as a symposium for the First International Congress of Systematic and Evolutionary Biology in Boulder, Colorado, during 1973. From the ten contributors, at least half of the papers deal with insect-plant coevolution and interactions. One paper in particular, written by Gordon W. Frankie, is directly applicable to rainforests: *Tropical Forest Phenology and Pollinator Plant Coevolution.*

TROPICAL ECOLOGICAL SYSTEMS - TRENDS IN TERRESTRIAL AND AQUATIC RESEARCH
Golley, Frank B. and Ernesto Medina (Editors)
1975. Springer-Verlag

This book presents 25 papers from 42 contributors that in 1973 attended the second meeting of the International Society of Tropical Ecology (ISTE) and the International Association for Ecology (INTECOL). Papers are grouped into the following sections: Physiological Ecology, Dynamics of Populations, Interaction Between Species, Tropical Forest Analysis, Savannas, Tropical Water Bodies, Island Ecosystems, and Applications.

TROPICAL RAIN FOREST ECOSYSTEMS - STRUCTURE AND FUNCTION
Golley, Frank B. (Editor)
1983. Elsevier Scientific Publishing Company, Inc.

This book is number 14A of the Ecosystems Of The World series. It contains 21 contributors who have written chapters under the following broad subject headings: Forest Structure, Forest Function, Physiological and Behavioral Aspects of Tropical Forest Biology, and Applications and Human Use of Forests.

TROPICAL TREES AND FORESTS - AN ARCHITECTURAL ANALYSIS
Halle, F., R.A.A. Oldeman and P.B. Tomlinson
1978. Springer-Verlag

The authors provide an analysis of tropical forests in terms of individual trees, seen as active, adaptable, interacting units. Much of the book describes tree 'architecture' (form and history of the form) and developmental models, which are used to describe tree strategies. Literature citations are given throughout the text.

VERTEBRATES IN COMPLEX TROPICAL SYSTEMS
Harmelin-Vivien, Mireille L. and Francois Bourliere (Editors)
1989. Springer-Verlag

This book originated from a symposium held at the Fourth International Congress of Ecology during 1986. The manuscripts presented discuss the factors affecting species richness in tropical animals. This was done by focusing on major vertebrate communities in rainforests and coral reefs. Chapters directly related to tropical forests are: Tropical Herpetofauna Communities: Patterns Of Community Structure In Neotropical Rainforests, Bird Community Structure in Two Rainforests: Africa (Gabon) and South America (French Guiana) - A Comparison, Mammalian Species Richness In Tropical Rainforests, and Species Diversity In Tropical Vertebrates: An Ecosystem Perspective.

THE FRAGMENTED FOREST - ISLAND BIOGEOGRAPHY THEORY AND THE PRESERVATION OF BIOTIC DIVERSITY
Harris, Larry D.
1984. University of Chicago Press

Harris defines the purpose of his book "to draw together available scientific information from the western Cascades and use it to evaluate the utility of island biogeography theory as a guide to comprehensive planning for the conservation of old-growth ecosystems in the context of managed forest lands." Treating old-growth forest areas as islands, the author outlines a plan for preserving biotic diversity by surrounding old-growth forest 'cores' with buffer zones of forest under long-rotation management. Although the data used by Harris to support his theory are taken from fieldwork conducted in the Pacific Northwest (USA), the ideas are applicable to old-growth forests in tropical environments.

FOREST ENVIRONMENTS IN TROPICAL LIFE ZONES - A PILOT STUDY
Holdridge, L.R., W.C. Grenke, W.H. Hatheway, T. Liang and J.A. Tosi, Jr.
1971. Pergamon Press

Based on the abstract of this book, it is a report that examines the potential of the Holdridge Classification of World Life Zones "to organize quantitative data on undercanopy environments into a predictive system".

TROPICAL FORESTS: BOTANICAL DYNAMICS, SPECIATION AND DIVERSITY
Holm-Nielson, L.B., H. Balsley and I. Nelson (Editors)
1989. Academic Press

This work presents the proceedings of a conference held at the University of Aarhus in Denmark during August 1988.

LA SELVA NUBLADA DE RANCHO GRANDE - PARQUE NACIONAL "HENRI PITTIER" - EL AMBIENTE FÍSICO, ECOLOGÍA Y ANATOMÍA VEGETAL
Huber, Otto (Editor)
1986. Fondo Editorial - Acta Cientifica Venezolana, Seguros Anauco C.A. (Caracas)

This excellent work presents a wealth of information about Rancho Grande under the general headings mentioned in the title. The research of 11 contributors is presented in 10 chapters dealing with the geography, climate, soils, and numerous aspects of the plant life that occurs in the cloud forest. Illustrations are provided in the form of line drawings, black and white photographs, and several color plates.

THE TROPICAL RAIN FOREST - A FIRST ENCOUNTER
Jacobs, Marius
1988. Springer-Verlag

This book was translated into English from the original Dutch version that was published in 1981. It is an excellent work for the serious layman or biologist. It treats both biotic and abiotic aspects of tropical forests and discusses rainforests throughout the world. It also deals with primary and secondary forests, forest communities that border rainforests, the interaction of man and the rainforest, and a chapter on how rainforests are studied.

ECOLOGY OF PLANTS IN THE TROPICS
Janzen, Daniel H.
1975. Edward Arnold Ltd.

This short book is number 58 in the Studies In Biology series put out by the Institute of Biology (London). In it, Janzen discusses tropical angiosperms and various aspects of their biology and ecology. The following topics are presented by chapters: Vegetative Biology, Pollination Biology, Fruit and Seed Biology, Chemical Defenses, Community Structure, Tropical Agriculture, and Suggested Field Studies.

COSTA RICAN NATURAL HISTORY
Janzen, Daniel H. (Editor)
1983. University of Chicago Press

Based on the research and experience of 174 contributors, an extensive amount of data on both the biotic and abiotic features of Costa Rica are presented. Checklists and species accounts are given for trees, insects, reptiles and amphibians, birds, and mammals. Much of the information is applicable to other Latin American countries.

TROPICAL ECOLOGY
Jordan, Carl F. (Editor)
1981. Hutchinson Ross Publishing Company

This book is number 10 in the series Benchmark Papers In Ecology. It presents 27 articles taken from refereed scientific journals and books, dividing them into two parts: Species Richness of the Tropics, and Functioning of Tropical Ecosystems. The editor comments on each of the articles presented.

NUTRIENT CYCLING IN TROPICAL FOREST ECOSYSTEMS - PRINCIPLES AND THEIR APPLICATION IN MANAGEMENT AND CONSERVATION
Jordan, Carl F.
1985. John Wiley and Sons

In six chapters with conclusions, Jordan discusses factors controlling nutrient cycles, nutrient conserving mechanisms, differences in ecosystem characteristics along environmental gradients, the characterization of nutrient cycles, and the changes in nutrient cycles due to disturbance. Jordan states in the Preface the objective of the text "...is to develop an understanding of how and why tropical forests are different from temperate zone forests, of the factors important in sustaining productivity of tropical forests, pastures, cropland and plantations, and of how this knowledge can contribute to better management of tropical ecosystems".

AMAZONIAN RAIN FORESTS: ECOSYSTEM DISTURBANCE AND RECOVERY: CASE STUDIES OF ECOSYSTEM DYNAMICS UNDER A SPECTRUM OF LAND USE-INTENSITIES
Jordan, Carl F. (Editor)
1987.

The articles in this book analyze and compare a variety of land uses and development strategies in the Amazon region. Case studies are presented from Brazil, Venezuela, and Peru.

TROPICAL BOTANY
Larsen, Kai and Lauritz B. Holm-Nielsen (Editors)
1979. Academic Press

This book contains the proceedings of a symposium held at the University of Aarhus (Denmark) in August 1978. Thirty authors have contributed as many papers under the following broad sections: History of Tropical Floras, General Phytogeography of Tropical Floras, Regional Phytogeography and Investigation of Tropical Floras, and Taxonomic and Biological examples. Much attention is given to neotropical species under each section.

THE UPPER AMAZON
Lathrap, Donald W.
1970. Thames and Hudson

This is Volume 73 in a series titled Ancient Peoples And Places. The author discusses tropical forest cultures, cultural development and ethnography (especially in the Ucayali Basin) in the upper Amazon.

THE ECOLOGY OF A TROPICAL FOREST - SEASONAL RHYTHMS AND LONG-TERM CHANGES
Leigh, Egbert G. Jr., A. Stanley Rand and Donald M. Windsor
1982. Smithsonian Institution Press

This work is a collection of scientific papers from research conducted at the Smithsonain Tropical Research Institute on Barro Colorado Island in Panama. This rainforest is probably the best studied in the New World, which is reflected in the 32 papers presented in this volume. Articles are grouped according to the following sections: the Physical Setting, the Biotic Setting, Seasonal Rhythms in Plants, Frugivores, Insects of Tree Crowns and Their Predators, Litter Arthropods and Their Predators, and Long-Term Changes.

TROPICAL FOREST AND ITS ENVIRONMENT
Longman, Kenneth A. and Jan Jenik
1987. Longman Scientific and Technical

This is a textbook that draws on research that has been conducted throughout the world, with slightly more emphasis on the region of West Africa. The book is divided into the following sections: Some Misconceptions, Forest and Environment Interacting, Environmental Factors, the Forest Community, Tree Growth Physiology, Dynamic Forest Ecosystems, and Management of Tropical Forest Land.

SPECIATION IN TROPICAL ENVIRONMENTS
Lowe-McConnel, R.H.
1969. Academic Press Inc.

This book is a collection of 15 papers that were given at a symposium on Speciation In Tropical Environments held in London during 1968. This symposium was organized jointly by the Linnean Society of London and the Tropical Group of the British Ecological Society. Speciation of various plant and animal groups from different tropical areas throughout the world is discussed.

PEOPLE AND THE TROPICAL FOREST
Lugo, Ariel E., John J. Ewel, Susanna B. Hecht, Peter G. Murphy, Christine Padoch, Marianne C. Schmink, and Donald Stone (Editors)
1987. U.S. Government Printing Office

This work contains 20 research reports that have resulted from projects conducted worldwide through the U.S. Man and the Biosphere Program. All these projects deal with the management of tropical forests in specific localities. The research summaries are categorized and presented under the following sections: Human Ecology and Agroforestry, General Ecology, and Forestry.

TROPICAL RAIN FOREST ECOLOGY
Mabberley, D.J.
1983. Blackie (Distributed in the U.S. by Chapman and Hall)

This work was written as a textbook intended for advanced undergraduate students. Among the topics covered are succession, biotic and abiotic factors, diversity and species richness, coevolution, and current uses of rain forests. Literature citations are included in the text, which discusses forests in Africa, Asia, and the Neotropics. There are few black and white photos and line drawings.

TROPICAL FOREST ECOSYSTEMS IN AFRICA AND SOUTH AMERICA: A COMPARATIVE REVIEW
Meggers, Betty J., Edward S. Ayensu and W. Donald Duckworth
1973. Smithsonian Institution Press

This volume contains 25 papers that were to be presented at a symposium (which was cancelled), planned by the Association for Tropical Biology. The flora and fauna of lowland tropical forest ecosystems of South America and Africa are compared, with special attention to examples of convergence and adaptive problems.

REACHING THE RAINFOREST ROOF - A HANDBOOK ON TECHNIQUES OF ACCESS AND STUDY IN THE CANOPY
Mitchell, Andrew W.
1982. Leeds Philosophical and Literary Society / United Nations Environment Programme

This booklet describes various methods currently available for scaling and studying the upper reaches of the rainforest canopy. Techniques are described in detail and illustrated with both drawings and photographs. Among the procedures discussed are climbing, towers, platforms, rope webs, and aerial walkways.

THE DISPERSAL CENTRES OF TERRESTRIAL VERTEBRATES IN THE NEOTROPICAL REALM: A STUDY IN THE EVOLUTION OF THE NEOTROPICAL BIOTA AND ITS NATIVE LANDSCAPES
Müller, Paul
1973. Junk Publishers

This book is actually a dissertation that was presented to the University of Saarland, Saarbrucken (1970). The author attempts to establish the existence of neotropical dispersal centers in an attempt to establish geographical subdivisions in Central and South America, based on faunal analyses. Forty Latin American dispersal centers are identified and discussed.

CONVERSION OF TROPICAL MOIST FORESTS
Myers, Norman
1980. National Academy of Sciences

This is a report that was prepared by the author for the Committee on Research Priorities in Tropical Biology of the National Research Council. It presents the results of a survey conducted during 1978-79, to establish the status and rates of conversion of tropical moist forests around the world. The roles of farmers, the timber trade, cattle production, and firewood cutting are all examined with respect to affecting conversion rates. Regional reviews of tropical Latin America, Africa, and southeast Asia are provided.

THE PRESERVATION OF SPECIES - THE VALUE OF BIOLOGICAL DIVERSITY
Norton, Bryan G. (Editor)
1986. Princeton University Press

This book was written under the auspices of the University of Maryland's Center for Philosophy and Public Policy. Its purpose was to analyze and interpret existing scientific data and to present them with analyses and possible solutions to the problem. Each of the 11 chapters are written by a separate author and grouped under the three broad sections titled: The Problem, Values And Objectives, and Management Considerations.

A TROPICAL RAIN FOREST - A STUDY OF IRRADIATION AND ECOLOGY AT EL VERDE, PUERTO RICO
Odum, Howard T. (Editor)
1970. Office of Information Services, U.S. Atomic Energy Commission

This three book series details a research project performed by the U.S. Atomic Energy Commission to determine the effects of gamma radiation on a rainforest site. The first book describes the project, the rainforest site, and the radiation experiment. The second contains information about the effects of radiation on plants, animals and microorganisms. The third book discusses cytological studies, soils and mineral cycling, and forest metabolism and energy flows.

BIOLOGICAL DIVERSIFICATION IN THE TROPICS
Prance, Ghillean T. (Editor)
1982. Columbia University Press

This book is concerned with the distribution and evolution of the species of tropical lowland forest in relation to the refuge theory. It presents most of the papers given at the fifth symposium of the Association for Tropical Biology. A total of 37 articles by 41 contributors discuss refuge theory in relation to the distribution of various taxa of flora and fauna throughout the world, but with emphasis on South America.

TROPICAL RAIN FORESTS AND THE WORLD ATMOSPHERE
Prance, Ghillean T. (Editor)
1986. Westview Press

This book was based on a symposium held in New York in 1984 by the Amercian Association for the Advancement of Science (AAAS), and is part of the AAAS Selected Symposia Series. The 13 contributors to this volume discuss tropical rainforests, their depletion, their interactions with the atmosphere, and potential effects of deforestation. Strong coverage is given to Amazonia.

EXTINCTION IS FOREVER: THREATENED AND ENDANGERED SPECIES OF PLANTS IN THE AMERICAS AND THEIR SIGNIFICANCE IN THE FUTURE
Prance, Ghillean T. and Thomas S. Elias (Editors)
1977. New York Botanical Garden

This book is the proceedings of a symposium held at the New York Botanical Garden in 1976. At least half of the text, including 13 individual papers, is devoted to the endangered flora of Central and South America.

AMAZONIA
Prance, Ghillean T. and Thomas E. Lovejoy (Editors)
1985. Pergamon Press

This volume is part of the Key Environments series produced in collaboration with the International Union For Conservation Of Natural Resources. It presents 22 papers from 23 contributors, all well known for their research in the Amazon. Articles are broadly divided into three sections: the Physical Setting, the Biology, and the Human Impact. The book's purpose was to collect and summarize knowledge on the Amazon environment and its flora and fauna.

THE TROPICAL RAIN FOREST
Richards, P.W.
1952. Cambridge University Press

This is considered the original classical work on rainforests, and although reprinted many times, it has not been substantially revised. It does not discuss fauna, but concentrates on describing plants and trees, their communities, succession, and the biotic and abiotic factors affecting them.

CONSERVATION BIOLOGY - THE SCIENCE OF SCARCITY AND DIVERSITY
Soule', Michael E. (Editor)
1986. Sinauer Associates, Inc.

This book presents 25 articles by 45 scientists discussing aspects of conservation biology, much of which is based on their own research. The contributions are grouped in the following sections: The Fitness And Viability Of Populations, Patterns Of Diversity And Rarity: Their Implications For Conservation, The Effects Of Fragmentation, Community Processes, Sensitive Habitats: Threats And Management, and Interacting With The Real World. At least one fourth of the articles deal with neotropical forest or plant communities, fauna, and/or deforestation.

CLOUD FOREST IN THE HUMID TROPICS
Stadtmüller, Thomas
1987. INFORAT, CATIE (Turrialba, Costa Rica)

This book is available in both English and Spanish. This short book describes the ecology of cloud forests in the humid tropics. In addition to discussing the extension of cloud forests and defining the term itself, the author treats their structure, composition, and silvicultural aspects, as well as their conversion and conservation. There is an excellent bibliography of Spanish, English, and German works.

TROPICAL RAIN FOREST: ECOLOGY AND MANAGEMENT
Sutton, Stephen L., T.C. Whitmore and A.C. Chadwick (Editors)
1983. Blackwell Scientific Publications

This book contains the papers presented at a symposium in 1982 that was sponsored by the Leeds Philosophical and Literary Society and the British Ecological Society. A total of 34 papers have been collected from some of the most distinguished names in tropical biology. Broad subject areas discussed include aspects of community structure and diversity, plant/animal interactions, decomposition and nutrient cycling, and resource management.

FIVE NEW WORLD PRIMATES - A STUDY IN COMPARATIVE ECOLOGY
Terborgh, John
1983. Princeton University Press

This book describes the behavior, activities, and diet of monkey species that were studied over a period of years. It is the first in a series called Monographs In Behavior And Ecology. It is based on research and observations conducted by the author and his associates near their field camp and research station at Cocha Cashu in Manu National Park of southeast Peru.

TROPICAL FOREST ECOSYSTEMS
UNESCO/UNEP/FAO
1978. UNESCO

This extensive 'state-of-knowledge' report was prepared by the United Nations Educational, Scientific and Cultural Organization (UNESCO) in conjunction with the United Nations Environment Programme (UNEP), and the Food and Agriculture Organization (FAO). Its purpose was "to provide a clear summary of knowledge of the structure, functioning and evolution of tropical forest ecosystems, and of the human populations that live within and around these ecosystems". The contents are divided into three main sections: Description, Functioning and Evolution of Tropical Forest Ecosystems, Man and the Patterns of Use of Tropical Forest Ecosystems, and Regional Case Studies. A portion of the latter section is devoted to the forest ecosystems of the Brazilian Amazon.

ECOLOGY OF TROPICAL PLANTS
Vickery, Margaret L.
1984. John Wiley and Sons

This book is a text, adapted for tropical regions from *PLANTS AND ENVIRONMENT - A TEXTBOOK OF AUTECOLOGY (3/e)* by R.F. Daubenmire. It treats the interaction of plants with each aspect of their physical and organic environment, including soil, water, radiation, atmosphere, animals, man, and other plants.

BIOGEOGRAPHY AND QUATERNARY HISTORY IN TROPICAL AMERICA
Whitmore, T.C. and G.T. Prance (Editors)
1987. Oxford University Press

This text attempts to analyze and appraise the theory of tropical evolution in paleoecological refugia. Contributors analyze distribution patterns for plants, birds, and butterflies, while the early history of man in Amazonia is also discussed. A final section is presented with conclusions and alternative hypotheses.

BIODIVERSITY
Wilson, Edward O. (Editor)
1988. National Academy Press

This book is a collection of 57 papers resulting from the National Forum on BioDiversity held in 1986, under the auspices of the Smithsonian Institution and the National Academy of Sciences. It discusses the major aspects of biodiversity, including its importance to man and its relation to tropical forest destruction. The papers presented are by scientists of many disciplines, describing research results from throughout the world.

A WORLD CENSUS OF TROPICAL ECOLOGISTS
Yantko, Joan A. and Frank B. Golley (Editors)
1977. Institute of Ecology, University of Georgia

This census provides a list of international tropical ecologists, including research workers, administrators, teachers, and environmental managers. The information listed for each person includes name, address, research specialty, and region of interest. Although much information has probably become outdated in the past 13 years, this census remains a useful and informative communications tool.

FLORA AND FAUNA GUIDES

During the late 1800s, a series of books were published on the results of a number of scientific expeditions by British researchers in Central America. The title is in Latin (BIOLOGIA CENTRALI AMERICANA); the majority of the text is in English, but with a few exceptions. These volumes cover invertebrates including arachnids, insects, and mollusks, as well as mammals, birds, and fish. There are also volumes on botany, zoology, and archaeology.

Birds

MANUAL OF NEOTROPICAL BIRDS - VOLUME 1
Blake, Emmet R.
1977. University of Chicago Press

BIRDS OF THE WEST INDIES
Bond, James
1985. Houghton Mifflin Co.

A FIELD GUIDE TO THE BIRDS OF MEXICO AND CENTRAL AMERICA
Davis, L. Irby
1972. University of Texas Press

SOUTH AMERICAN LAND BIRDS: A PHOTOGRAPHIC AID TO IDENTIFICATION
Dunning, John S. (Edited by Robert S. Ridgely)
1982. Harrowood Books

SOUTH AMERICAN BIRDS: A PHOTOGRAPHIC AID TO IDENTIFICATION
Dunning, John S.
1989. Harrowood Books

A FIELD GUIDE TO THE BIRDS OF MEXICO
Edwards, Ernest P.
1989. Ernest P. Edwards (Sweet Briar, VA)

A GUIDE TO THE BIRDS OF TRINIDAD AND TOBAGO
ffrench, Richard
1980. Harrowood Books

AVES BRASILEIRAS
Frisch, J.D.
1981. Dalgas-Ecologia Tecnica e Comercio Ltda.

FIELD GUIDE TO THE BIRDS OF THE GALAPAGOS
Harris, Michael
1982. Collins (London)

A MANUAL FOR BIRD WATCHING IN THE AMERICAS
Heintzelman, Donald S.
1979. Universe Books

A GUIDE TO THE BIRDS OF COLOMBIA
Hilty, Steven and William L. Brown
1986. Priceton University Press

BIRDS OF GUATEMALA
Land, Hugh C.
1970. Livingston Publishing Co.

A GUIDE TO THE BIRDS OF SOUTH AMERICA
Meyer de Schauensee, Rodolphe
1970. Academy of Natural Sciences of Philadelphia

A GUIDE TO THE BIRDS OF VENEZUELA
Meyer de Schauensee, R. and W.H. Phelps, Jr.
1978. Princeton University Press

A FIELD GUIDE TO MEXICAN BIRDS; FIELD MARKS OF ALL SPECIES FOUND IN MEXICO, GUATEMALA, BELIZE (BRITISH HONDURAS), EL SALVADOR
Peterson, Roger T. and Edward L. Chalif
1973. Houghton Mifflin Co.

A GUIDE TO THE BIRDS OF PANAMA
Ridgely, Robert S.
1981. Princeton University Press

A GUIDE TO THE BIRDS OF PANAMA, WITH COSTA RICA, NICARAGUA, AND HONDURAS
Ridgely, Robert S. and John Gwynne, Jr.
1989. Princeton University Press

THE BIRDS OF SOUTH AMERICA - THE OSCINE PASSERINES
Ridgely, Robert S. and Guy Tudor
1989. University of Texas Press

A GUIDE TO THE BIRDS OF COSTA RICA
Stiles, Gary and Alexander F. Skutch
1989. Cornell University Press

THE BIRDS OF THE REPUBLIC OF PANAMA
Wetmore, Alexander
1965. Smithsonian Institution

Several other neotropical bird guides are in the process of being written. *THE BIRDS OF ECUADOR* with text by Robert S. Ridgely and plates by Paul Greenfield is tentatively scheduled for publication in 1993. A guide to the birds of Peru is also in the works with Ted Parker III as one of the authors.

Butterflies

BUTTERFLIES OF TRINIDAD AND TOBAGO
Barcant, Malcolm
1970. Collins Clear-Type Press

BUTTERFLIES OF SOUTH AMERICA
D' Abrera, Bernard
1984. Hill House

BUTTERFLIES OF THE NEOTROPICAL REGION   PART 1.
PAPILIONIDAE AND PIERIDAE
D' Abrera, Bernard
1981. Lansdowne Editions (Distributed by E.W. Classey)

BUTTERFLIES OF THE NEOTROPICAL REGION   PART 2.
DANAIDAE, ITHOMIIDAE, HELICONIDAE, AND MORPHIDAE
D' Abrera, Bernard
1984. Hill House

MARIPOSAS DIURNAS DE VENEZUELA - INTODUCCION A SU CONOCIMIENTO
Alvarez Sierra, Jose Ramon and Jose Ramon Alvarez Corral
1984. Caracas

THE BUTTERFLIES OF COSTA RICA AND THEIR NATURAL HISTORY
DeVries, Philip J.
1987. Princeton University Press

BUTTERFLIES
Emmel, Thomas C.
1975. Alfred A. Knopf

LES MORPHO D' AMERIQUE DU SUD ET CENTRALE: HISTORIQUE - MORPHOLOGIE - SYSTEMATIQUE
Le Moult, E.
1962. Editions du Cabinet Entomologique

BUTTERFLIES OF THE WORLD
Preston-Mafham, Rod and Ken Preston-Mafham
1988. Facts-On-File Publications

A FIELD GUIDE TO THE BUTTERFLIES OF THE WEST INDIES
Riley, Norman D.
1975. William Collins Sons

MARIPOSAS DE VENEZUELA
Schmid, Michael and Bradford M. Endicott
1968. L. Levinson Junr. Ltd (Denmark)

Mammals

MAMMALS OF THE NEOTROPICS - VOLUME 1 THE NORTHERN NEOTROPICS - PANAMA, COLOMBIA, VENEZUELA, GUYANA, SURINAME, FRENCH GUIANA
Eisenberg, John F.
1989. University of Chicago Press

NEOTROPICAL RAINFOREST MAMMALS - A FIELD GUIDE
Emmons, Louise H.
(To be published in 1990). University of Chicago Press

THE MAMMALS OF NORTH AMERICA (Two Volumes)
Hall, E. Raymond
1981. John Wiley

LIVING NEW WORLD MONKEYS (PLATYRRHINI) - VOLUME 1
Hershkovitz, Philip
1977. University of Chicago Press

THE MAMMALS OF SURINAME
Husson, A.M.
1978. E.J. Brill (Leiden)

MURCIELAGOS DE VENEZUELA
Linares, Omar J.
Cuadernos Lagoven

LOS PRINCIPALES MAMIFEROS SILVESTRES DE PANAMA
Mendez, Eustorgio
1970. Panama

THE NEW WORLD PRIMATES - ADAPTIVE RADIATION AND THE EVOLUTION OF SOCIAL BEHAVIOR, LANGUAGES, AND INTELLIGENCE
Moynihan, Martin
1976. Princeton University Press

FIVE NEW WORLD PRIMATES - A STUDY IN COMPARATIVE ECOLOGY
Terborgh, John
1983. Princeton University Press

MAMMALS OF THE WORLD
Walker, Ernest P.
1975. John Hopkins University Press

Reptiles And Amphibians

BRAZILIAN SNAKES: A COLOR ICONOGRAPHY
Amaral, Afranio do
1977. Edicoes Melhoramentos (Sao Paulo)

VENOMOUS REPTILES OF LATIN AMERICA
Campbell, Jonathan A. and William W. Lamar
1989. Cornell University Press

FROGS OF SOUTHEASTERN BRAZIL
Cochran, Doris M.
1955. Bulletin U.S. National Museum. 206: 1-423.

FROGS OF COLOMBIA
Cochran, Doris M. and Coleman J. Goin
1970. Bulletin U.S. National Museum. 288: 1-655.

OFIDIOS DA AMAZONIA
da Cunha, Osvaldo R. and Francisco P. do Nascimento
1978. Museu Paraense Emilio Goeldi, Publicacoes Arulsas No. 31

THE REPTILES OF THE UPPER AMAZON BASIN, IQUITOS REGION, PERU
Dixon, James R. and Pekka Soini
1986. Milwaukee Public Museum

HYLID FROGS OF MIDDLE AMERICA
Duellman, William E.
1970. Monograph of the Museum of Natural History, University of Kansas

THE BIOLOGY OF AN EQUATORIAL HERPETOFAUNA IN AMAZONIAN ECUADOR
Duellman, William E.
1978. University of Kansas Museum of Natural History, Miscellaneous Publication 65

SOUTH AMERICAN HERPETOFAUNA: ITS ORIGIN, EVOLUTION AND DISPERSAL
Duellman, William E. (Editor)
1979. Kansas Museum of Natural History Monograph: No. 7. University of Kansas Museum of Natural History

THE WORLD OF VENOMOUS ANIMALS
Freiberg, Marcos A. and Jerry G. Walls
1984. TFH Publications

NOTES ON THE HERPETOFAUNA OF SURINAM IV. THE LIZARDS AND AMPHISBANIANS OF SURINAM
Hoogmoed, M.S.
1973. Biogeographica, Vol. IV. D.R. Junk, Publishers, The Hague

LIVING SNAKES OF THE WORLD IN COLOR
Mehrtens, John M.
1987. Sterling Publishing Co.

*CATALOGUE OF THE NEOTROPICAL SQUAMATA: PART 1.
SNAKES, PART 2. LIZARDS AND AMPHIBIANS*
Peters, James A., Roberto Donoso-Barros and Braulio Orejas-Miranda
1986. Smithsonian Institution Press

INTRODUCTION TO THE HERPETOFAUNA OF COSTA RICA
(Text in English and Spanish)
Savage, Jay M. and Jaime Villa R.
1986. Society for the Study of Amphibians and Reptiles

GUIDE TO THE IDENTIFICATION OF THE AMPHIBIANS AND REPTILES OF THE WEST INDIES (EXCLUSIVE OF HISPANIOLA)
Schwartz, Albert and Robert W. Henderson
1985. Milwaukee Public Museum

Floras

GEOGRAPHICAL GUIDE TO THE FLORAS OF THE WORLD - AN ANNOTATED LIST WITH SPECIAL REFERENCE TO USEFUL PLANTS AND COMMON PLANT NAMES
Blake, S.F. and A.C. Atwood
USDA Miscellaneous Publication No. 401
1942. U.S. Government Printing Office

FLORA COSTARICENSIS
Burger, William (Editor)
Fieldiana: Botany
Field Museum of Natural History (Chicago)

FLORISTIC INVENTORY OF TROPICAL COUNTRIES - THE STATUS OF PLANT SYSTEMATICS, COLLECTION, AND VEGETATION PLUS RECOMMENDATIONS FOR THE FUTURE
Campbell, David G. and H. David Hammond (Editors)
1989. The New York Botanical Garden

TROPICAL TIMBERS OF THE WORLD
Chudnoff, Martin
Agriculture Handbook: No. 607
1984. U.S. Government Printing Office

TROPICALS
Courtright, Gordon
1988. Timber Press

FLORA OF BARRO COLORADO ISLAND
Croat, Thomas B.
1978. Stanford University Press

FLORA OF PANAMA
D' Arcy, W.G.
Annals of the Missouri Botanical Garden
1959-Present

FLORA OF THE RIO PALENQUE SCIENCE CENTER, LOS RIOS, ECUADOR
Dodson, C.H. and A.H. Gentry
1978. Selbyana 4: 1-628.

LA FLORA DE JAUNECHE, LOS RIOS, ECUADOR
Florulas de las Zonas de Vida del Ecuador
Dodson, C.H., A.H. Gentry and F.M. Valverde
1985. Banco Central del Ecuador, Quito

FLORA OF GUATEMALA
Fieldiana: Botany
Field Museum of Natural History (Chicago)

FLORA OF THE GUIANAS
Gorts-van Rijn, A.R.A. (Editor)
Koeltz Scientific Books. 1985-1986-1987-1988

FLORA OF ECUADOR
Harling, Gunnar and Lennart Andersson (Editors)
Department of Systematic Botany, University of Goteborg, and Section for Botany
Riksmuseum, Stockholm, Sweden   1986 and 1987

FLORA DE VENEZUELA
Lasser, Tobias
Edicion Especial Del Instituto Botanico

THE COLLINS GUIDE TO TROPICAL PLANTS
Lotschert, William and Gerhard Beese
1989. William Collins Ltd.

FLORA NEOTROPICA   (Monographs)
New York Botanical Garden
1974-Present

A FIELD GUIDE TO TROPICAL AND SUBTROPICAL PLANTS
Perry, Frances and Roy Hay
1982. Van Nostrand Reinhold Co.

STUDIES ON TROPICAL ANDEAN ECOSYSTEMS - VOLUME I LA CORDILLERA CENTRAL COLOMBIANA - TRANSECTO PARQUE LOS NEVADOS
Van Der Hammen, Thomas, Alfonso P. Preciado and Polidoro Pinto E. (Editors)
1983. J. Cramer (Germany)

STUDIES ON TROPICAL ANDEAN ECOSYSTEMS - VOLUME 2 LA SIERRA NEVADA DE SANTA MARTA (COLOMBIA) - TRANSECTO BURITACA - LA CUMBRE
Van Der Hammen, Thomas and Pedro M. Ruiz (Editors)
1984. J. Cramer (Germany)

FLORA OF PANAMA
Woodson, R.E., Jr.
Schery, Robert W. and collaborators

ENVIRONMENTAL, POLITICAL, AND MISCELLANEOUS TITLES

CENTRAL AMERICA, THE WEST INDIES AND SOUTH AMERICA
Bates, H.
1976. Gordon Press

UNDERDEVELOPING THE AMAZON - EXTRACTION, UNEQUAL EXCHANGE, AND THE FAILURE OF THE MODERN STATE
Bunker, Stephen G.
1985. University of Chicago Press

PROCEEDINGS OF THE U.S. STRATEGY CONFERENCE ON BIOLOGICAL DIVERSITY
Bureau of International Organization Affairs
1982. Department of State Publication 9262. International Organization and Conference Series 300

SAVING CRITICAL ECOSYSTEMS
Burley, F. William
1988. World Resources Institute

TROPICAL FOREST ACTION PLAN
Burley, F.W. and P. Hazlewood
1986. Annual Report of the World Resources Institute

REHABILITATING DAMAGED ECOSYSTEMS (2 VOL.S)
Cairns Jr., John (Editor)
1988. CRC Press

INDIGENOUS PEOPLES AND TROPICAL FORESTS: MODELS OF LAND USE AND MANAGEMENT FROM LATIN AMERICA
Clay, Jason W.
1988. Cultural Survival

TO FEED THE EARTH - AGRO-ECOLOGY FOR SUSTAINABLE DEVELOPMENT
Dover, Michael and Lee Talbot
1987. World Resources Institute

ECOLOGY, RECREATION AND TOURISM
Edington, John M. and M. Ann Edington
1986. Cambridge University Press

HUMAN CARRYING CAPACITY OF THE BRAZILIAN RAINFOREST
Fearnside, Phillip M.
1986. Columbia University Press

THE EARTH REPORT: THE ESSENTIAL GUIDE TO GLOBAL ECOLOGICAL ISSUES
Goldsmith, Edward and Nicholas Hilyard (Editors)
1988. Price Stern

AMAZON JUNGLE: GREEN HELL TO RED DESERT?
Goodland, R.J.A. and H.S. Irwin
1975. Elsevier Scientific Publishing Company

THE TROPICAL WORLD - ITS SOCIAL AND ECONOMIC CONDITIONS AND ITS FUTURE STATUS
Gourou, Pierre
1966. John Wiley and Sons, Inc.

HIGHWAYS INTO THE UPPER AMAZON BASIN - PIONEER LANDS IN SOUTHERN COLOMBIA, ECUADOR, AND NORTHERN PERU
Hegen, Edmund E.
1966. University of Florida Press

THE USE OF ECOLOGICAL GUIDELINES FOR DEVELOPMENT IN THE AMERICAN HUMID TROPICS
1975. International Union for Conservation of Nature and Natural Resources

LAND AND RESOURCE EVALUATION FOR NATIONAL PLANNING IN THE TROPICS
Lund, H. Gyde, Miguel Caballero-Deloya and Raul Villareal-Canton (Editors)
1987. General Technical Report WO-39, USDA, Forest Service

BORDERING ON TROUBLE - RESOURCES AND POLITICS IN LATIN AMERICA
Maguire, Andrew and Janet W. Brown (Editors)
1986. World Resources Institute

DEVELOPING THE AMAZON
Moran, E.
1981. Indiana University Press

THE PRIMARY SOURCE: TROPICAL FORESTS AND OUR FUTURE
Myers, Norman
1984. W.W. Norton

NOT FAR AFIELD: U.S. INTERESTS AND THE GLOBAL ENVIRONMENT
Myers, Norman
1987. World Resources Institute

ECOLOGICAL ASPECTS OF DEVELOPMENT IN THE HUMID TROPICS
Committee On Selected Biological Problems In The Humid Tropics (Chairman: Jay M. Savage)
National Academy of Sciences
1982. National Academy Press

RESEARCH PRIORITIES IN TROPICAL BIOLOGY
Committee On Research Priorities In Tropical Biology (Chairman: Peter H. Raven)
National Academy of Sciences
1980. National Academy Press

DECADE OF PROGRESS FOR SOUTH AMERICAN NATIONAL PARKS 1974-1984
National Park Service
U.S. Department of the Interior

WHY PRESERVE NATURAL VARIETY
Norton, Bryan G.
1987. Princeton University Press

ECOLOGICAL KNOWLEDGE AND ENVIRONMENTAL PROBLEM-SOLVING
Orians, Gordon H. (Editor)
1986. National Academy Press

DOWN TO BUSINESS - MULTINATIONAL CORPORATIONS, THE ENVIRONMENT, AND DEVELOPMENT
Pearson, Charles S.
1985. World Resources Institute

MULTINATIONAL CORPORATIONS, ENVIRONMENT, AND THE THIRD WORLD
Pearson, Charles S.
1987. World Resources Institute

SLASH AND BURN - FARMING IN THE THIRD WORLD FOREST
Peters, William J. and Leon F. Neuenschwander
1985. University of Idaho Press

PUBLIC POLICIES AND THE MISUSE OF FOREST RESOURCES
Repetto, Robert and Malcolm Gillis (Editors)
1988. Cambridge University Press

BIOLOGICAL PRIORITIES FOR CONSERVATION IN THE TROPICAL ANDES
Saavedra, Carlos and Curtis H. Freese
World Wildlife Fund - U.S.

FRONTIER EXPANSION IN AMAZONIA
Schmink, M. and Charles H. Wood (Editors)
1984. University of Florida Press

HOOFPRINTS ON THE FOREST: CATTLE RANCHING AND THE DESTRUCTION OF LATIN AMERICA'S TROPICAL FORESTS
Shane, D.R.
1986. Institute for the Study of Human Issues (Philadelphia)

BANKROLLING DISASTERS: INTERNATIONAL DEVELOPMENT BANKS AND THE GLOBAL ENVIRONMENT
Schwartzman, Stephan
1986. Sierra Club

RAINFOREST CORRIDORS: THE TRANSAMAZON COLINIZATION SCHEME
Smith, Nigel J.H.
1982. University of California Press

RESEARCH PRIORITIES FOR CONSERVATION BIOLOGY
Soule', Michael E. and Kathryn A. Kohm (Editors)
1989. Island Press

HELPING DEVELOPING COUNTRIES HELP THEMSELVES
Talbot, Lee
1985. World Resources Institute

TECHNOLOGIES TO SUSTAIN TROPICAL FOREST RESOURCES
Office of Technology Assessment
1984. U.S. Government Printing Office

TECHNOLOGIES TO MAINTAIN BIOLOGICAL DIVERSITY
Office of Technology Assessment
1987. U.S. Government Printing Office

TROPICAL DEFORESTATION - AN OVERVIEW THE ROLE OF INTERNATIONAL ORGANIZATIONS, THE ROLE OF MULTINATIONAL CORPORATIONS
1981. U.S. Government Printing Office

CONSERVATION, SCIENCE AND SOCIETY
United Nations, Educational, Scientific and Cultural Organization/United Nations Environment Programme
1984. United Nations Educational, Scientific and Cultural Organization

CONSERVATION FOR THE TWENTY-FIRST CENTURY
Western, David and Mary C. Pearl (Editors)
1989. Oxford University Press

TROPICAL FORESTS: A CALL FOR ACTION
World Resources Institute Staff
1985. Report of International Task Force convened by the World Resources Institute, the World Bank, and the United Nations Development Programme

THE GREENHOUSE TRAP
World Resources Institute
1990. Beacon Press

THE WORLD BANK AND AGRICULTURAL DEVELOPMENT - AN INSIDER'S VIEW
Yudelman, Montague
1985. World Resources Institute

The Forestry Private Enterprise Initiative has published a special series of 'working papers' through the Southeastern Center for Forest Economics Research. These papers cover a variety of topics, all related to forest enterprises, and many concerning nature-oriented tourism in the tropics. A free catalogue of papers available can be obtained at the following address:

Southeastern Center for Forest Economics Research
P.O. Box 12254
Research Triangle Park, North Carolina 27709 USA
Telephone: (919)-549-4030

The Board on Science and Technology for International Development (BOSTID) publishes a number of reports and documents based on programs it manages with developing countries. These publications deal with a variety of topics, including resource management, innovations in tropical forestry, managing tropical animal resources, and others. A catalog of BOSTID publications can be obtained by writing to the following address:

National Research Council
BOSTID Publications/HA-476E
2101 Constitution Ave., NW
Washington, D.C. 20418

The Island Press publishes an Annual Environmental Sourcebook. The 1990 edition lists "159 books for better conservation and management". Among the areas of interest covered by these books are tropical forests and sustainable development, biological diversity and wildlife, global warming, and sustainable agriculture. A copy of the Annual Environmental Sourcebook can be obtained by writing to:

Center For Resource Economics/Island Press
1718 Connecticut Avenue, N.W. Suite 300
Washington, D.C. 20009 USA

The organization Cultural Survival Inc. publishes and markets a variety of books dealing with indigenous peoples and cultural issues in developing countries around the world. More information about this organization and a catalogue of their titles can be obtained by writing to the following address:

Cultural Survival Inc.
11 Divinity Avenue
Cambridge, MA 02138 USA
Telephone: (617)-495-2562

TRAVEL GUIDES AND REGIONAL TITLES

The books listed in this section refer to South and Central America in general. Travel books referring to specific countries will be listed in the Books section following the description of that particular country and its rainforest sites.

BIRNBAUM'S SOUTH AMERICA 1989
Birnbaum, Stephen (Editor)
1988. Houghton Mifflin Company

SOUTH AMERICA: RIVER TRIPS
Bradt, George and Hilary Bradt
1982. Bradt Enterprises

1989 SOUTH AMERICAN HANDBOOK
Brooks, John (Editor)
1988. Trade and Travel Publications Ltd (Distributed in the U.S. by Prentice Hall)

RAINFORESTS: A GUIDE TO RESEARCH AND TOURIST FACILITIES AT SELECTED TROPICAL FOREST SITES IN CENTRAL AND SOUTH AMERICA
Castner, James L.
1990. Feline Press (P.O. Box 7219, Gainesville, FL 32605)

SOUTH AMERICA ON A SHOESTRING - INCLUDING MEXICO AND CENTRAL AMERICA
Crowther, Geoff
1986. Lonely Planet Publications

THE TRAVELER'S GUIDE TO LATIN AMERICAN CUSTOMS AND MANNERS
Devine, Elizabeth and Nancy L. Braganti
1988. St. Martin's Press

LETS EXPLORE CENTRAL AMERICA
Faber, Stuart J.
1975. Charing Cross

FROMMER'S SOUTH AMERICA ON $30 A DAY
Greenberg, Arnold and Harriet Greenberg
1987. Prentice Hall Press

SOUTH AMERICA TRAVEL DIGEST
Jacobs, Charles and Babette Jacobs
1986. Travel Digests

SOUTH AMERICAN RIVER TRIPS
Jordan, Tanis and Martin Jordan
1987. Bradt Publications (England)

MICHAEL'S GUIDE - SOUTH AMERICA
Shichor, Michael
1988. Inbal Travel Information Ltd (Israel)

FODOR'S CENTRAL AMERICA - EL SALVADOR, GUATEMALA, BELIZE, COSTA RICA, HONDURAS, NICARAGUA, PANAMA
Thompson, Alice (Editor)
1988. Fodor's Travel Publications, Inc.

FODOR'S 89 - SOUTH AMERICA - INCLUDING ANTARCTICA AND THE GALAPAGOS ISLANDS
Thompson, Alice (Editor)
1988. Fodor's Travel Publications, Inc.

NAGEL'S ENCYCLOPEDIA GUIDE: CENTRAL AMERICA (GUATEMALA, HONDURAS, BELIZE, SALVADOR, NICARAGUA, COSTA RICA, PANAMA)
1980. Nagel Publishers (Geneva)

A TRAVELER'S GUIDE TO EL DORADO AND THE INCAN EMPIRE
Meisch, Lynn
1984. Penguin Books

LATIN AMERICA ON A BICYCLE
Panet, J.S.
1987. Passport Press

STATISTICAL AND REFERENCE BOOKS/DIRECTORIES

THE WORLD OF LEARNING - 1989
1988. Europa Publications Limited (London)

THE EUROPA YEAR BOOK 1989
1989. Europa Publications Limited (London)

THE DIRECTORY OF MUSEUMS AND LIVING DISPLAYS
Hudson, Kenneth and Ann Nicholls
1985. Macmillan Publishers Limited (United Kingdom)

MUSEUMS OF THE WORLD
1981. K.G. Saur Verlag KG (Munich)

DIRECTORIO LATINOAMERICANO - SOCIO-ECONOMICO POLITICO ACADEMICO
Ediec Latina. Ediciones de informacion economica latinoamericana. Quito, Ecuador

A large number of reference books dealing with Latin America are published by Praeger Publishers and Greenwood Press, both of which are divisions of Greenwood Press, Inc.

Greenwood Press, Inc.
88 Post Road West
Box 5007
Westport, Connecticutt 06881 USA

The United States Government Printing Office publishes a number of books and booklets that are usually available at a reasonable price. Many of these publications can be extremely valuable to people intending to work in foreign countries. Examples of such publications are the *Diplomatic List, Key Officers of Foreign Service Posts,* and *Foreign Consular Offices In The United States.* The U.S. Department of State through its Bureau of Public Affairs publishes a series of informative papers called 'Background Notes'. These papers are usually several pages long and give a variety of general information about individual countries. They are available for almost all the countries that make up Latin America and can be ordered through the U.S. Government Printing Office.

U.S. Government Printing Office
Washington, D.C. 20402 USA
Telephone: (202)-783-3238

Many of the large zoological gardens, botanical gardens, and aquariums have curators and scientists actively involved in collecting and studying animals and plants indigenous to the Neotropics. Appendix D presents a selected list of some of the larger institutions of this nature in the United States. However, there are several excellent directories that give listings of zoological gardens, botanical gardens, and aquariums worldwide. The following are some of the most useful books for securing this type of information.

ZOOLOGICAL PARKS AND AQUARIUMS IN THE AMERICAS
Boyd, Linda (Editor)
1984-85. American Association of Zoological Parks and Aquariums

INTERNATIONAL ZOO YEARBOOK (Vol. 26)
Olney, P.J.S. (Editor)
1987. Zoological Society of London

THE WORLD OF ZOOS; A SURVEY AND GAZETTEER
Kirchshofer, Rosl (Compilor)
1968. Viking Press

INTERNATIONAL DIRECTORY OF BOTANICAL GARDENS IV
Henderson, D.M. (Compilor)
1983. Koeltz Scientific Books

INDEX HERBARIORUM - PART I THE HERBARIA OF THE WORLD (7th Edition)
Holmgren, Patricia K., Wil Keuken and Eileen K. Schofield
1981. Dr. W. Junk B.V., Publishers

MAGAZINES FOR NATURALISTS

BBC WILDLIFE
Broadcasting House
Whiteladies Road
Bristol BS8 2LR
England

BBC WILDLIFE is published jointly by BBC Magazines and Wildlife Publications Ltd. Articles treating conservation and wildlife are featured and illustrated with many color photographs.

EQUINOX
Telemedia Publishing (Publisher)
7 Queen Victoria Road
Camden East
Ontario K0K 1J0
Canada

EQUINOX is published bimonthly and covers all aspects of geography and science. Natural history articles frequently appear with excellent photography.

FOREST WORLD
World Forestry Center
4033 S.W. Canyon Road
Portland, OR 97221 USA

FOREST WORLD is published quarterly by the World Forestry Center for the education and enjoyment of readers who have an interest in the conservation of the world's forests. It features conservation-oriented articles of current interest illustrated with excellent color photography. The Fall 1989 issue contained several stories relating to tropical forests.

GEOMUNDO
Publicaciones Continentales De Mexico, S.A.
Lucio Blanco No 435
San Juan Tlihuaca

GEOMUNDO (continued)
Azcapotzalco 02400
Mexico, D.F.

GEOMUNDO Magazine is one of the few large circulation Spanish language magazines that deal with natural history topics. Articles are written for laymen and usually well illustrated with color photographs. GEOMUNDO is published monthly and distributed throughout Latin America. It can usually be found in bookstores in Latin America, or in kiosks on street corners.

INTERNATIONAL WILDLIFE
National Wildlife Federation (Publisher)
1412 Sixteenth Street, N.W.
Washington, D.C. 20036 USA

INTERNATIONAL WILDLIFE is a conservation magazine featuring high quality photography and published on a bimonthly basis. It is concerned with wildlife and ecology all over the world, as well as with all aspects of the environment.

NATIONAL GEOGRAPHIC
National Geographic Society (Publisher)
17th and M Streets, N.W.
Washington, D.C. 20036 USA

NATIONAL GEOGRAPHIC is a monthly magazine featuring exceptional color photography and directed toward an audience of well educated laymen. Articles dealing with particular countries may be located by searching through the index. An excellent story discussing rainforests can be found in the January 1983 (Vol. 163, No. 1) issue.

NATURAL HISTORY
American Museum of Natural History (Publisher)
Central Park West at 79th Street
New York, New York 10024 USA

NATURAL HISTORY is published monthly and contains popular natural history articles, usually based on original research by the author. Many color photographs typically illustrate the text.

ORION NATURE QUARTERLY
Myrin Institute
136 East 64th Street
New York, New York 10021 USA

ORION NATURE QUARTERLY is published four times a year by the Myrin Institute, and now in association with Conservation International (see page 295). One of its purposes is to encourage people to sustain biological diversity and the ecosystems and ecological processes that support life on earth. The magazine content reflects this theme with timely stories and is illustrated with excellent color photography.

ORYX
8-12 Camden High Street
London NW1 OJH
United Kingdom

ORYX is the quarterly journal published for the Fauna & Flora Preservation Society by Blackwell Scientific Publications Ltd. The articles in ORYX are easily read by layman or scientist and deal primarily with efforts being conducted to save endangered animal and plant species.

WILDLIFE CONSERVATION
New York Zoological Society
Bronx, New York 10460 USA

WILDLIFE CONSERVATION is currently published six times a year by the New York Zoological Society (NYZS). It often showcases the work of NYZS scientists, many of whom are actively involved in rainforest research projects. Prior to January 1990, *WILDLIFE CONSERVATION* was called *ANIMAL KINGDOM* magazine.

JOURNALS FOR BIOLOGISTS

BIOLOGICAL CONSERVATION
Elsevier Science Publishers Ltd.
Crown House
Linton Road
Barking, Essex 1G11 8JU
England

BIOLOGICAL CONSERVATION published in 1989 four volumes with four issues per volume. Its main purpose is to publish and disseminate information treating wildlife preservation and the conservation of biological natural resources.

BIOSCIENCE
American Institute of Biological Sciences (Publisher)
730 11th Street, NW
Washington, DC 20001-4584 USA

BIOSCIENCE is published monthly by the American Institute of Biological Sciences (AIBS). It is a refereed, cross-disciplinary journal for biologists. It publishes review articles and essays, as well as having regular special issues that focus on single topics.

BIOTROPICA
Association For Tropical Biology (Publisher)
Missouri Botanical Garden
P.O. Box 299
St. Louis, Missouri 63166 USA

BIOTROPICA is published quarterly and features refereed scientific manuscripts discussing all aspects of tropical biology.

BRENESIA
Museo Nacional de Costa Rica (Publisher)
Calle 17
Avenida Central y 2
Apartado 749
San Jose, Costa Rica

BRENESIA is the biological journal of the National Museum of Costa Rica.

ECOLOGY
Ecological Society of America
Center for Environmental Studies
Arizona State University
Tempe, Arizona 85287 USA

ECOLOGY is published six times a year by the Ecological Society of America. It publishes essays and articles that report and interpret the results of original scientific research concerned with the relationships between organisms and their environments.

ECOTROPICOS
Sociedad Venezolana de Ecologia (Publisher)
Apartado 47543
Caracas, Venezuela

ECOTROPICOS is a recent scientific journal that is published twice a year by the Venezuelan Society of Ecology.

ENVIRONMENTAL CONSERVATION
Elsevier Sequoia S.A.
P.O. Box 564
1001 Lausanne 1
Switzerland

ENVIRONMENTAL CONSERVATION is published quarterly by Elsevier Sequoia S.A. for the Foundation for Environmental Conservation. It is published with the collaboration of the United Nations Environment Programme (UNEP), the International Union for Conservation of Nature and Natural Resources (IUCN), the International Association for Ecology (INTECOL), the International Conferences on Environmental Future (ICEFs), the World Council For The Biosphere-International Society For Environmental Education (WCB-ISEE) and the World Wide Fund for Nature (WWF). *ENVIRONMENTAL CONSERVATION* is "devoted to maintaining global viability through exposing and countering environmental deterioration resulting from human population-pressure and unwise technology".

STUDIES ON NEOTROPICAL FAUNA AND ENVIRONMENT
Swets and Zeitlinger B.V. (Publisher)
Publishing Department
P.O. Box 825
2160 SZ Lisse
The Netherlands

This is a quarterly scientific journal that deals with the ecology, systematics, and distribution of the neotropical fauna. Earlier volumes were published under the title of *Beitrage zur neotropischen Fauna* (Volumes 1-6), and *Studies on the Neotropical Fauna* (Volumes 7-10).

JOURNAL OF DEVELOPING AREAS
Western Illinois University
900 West Adams Street
Macomb, IL 61455 USA

The *JOURNAL OF DEVELOPING AREAS* is published quarterly under the direction of Western Illinois University. Its purpose is to "stimulate the descriptive, theoretical, and comparative study of regional development, past and present, with the objective of promoting fuller understanding of man's relationship to the development process".

JOURNAL OF ECOLOGY
Blackwell Scientific Publications Ltd.
P.O. Box 88
Oxford OX2 0NE

The *JOURNAL OF ECOLOGY* is published four times per year by Blackwell Scientific Publications Ltd. for the British Ecological Society. It publishes original research papers on any aspect of plant ecology.

JOURNAL OF TROPICAL ECOLOGY
Cambridge University Press
32 East 57th Street
New York, New York 10022 USA

The *JOURNAL OF TROPICAL ECOLOGY* is published quarterly by Cambridge University Press for the International Association for Ecology (INTECOL) and the ICSU Press. It publishes papers dealing with any aspect of tropical ecology.

REVISTA DE BIOLOGÍA TROPICAL
Biblioteca de la Universidad de Costa Rica
Civdad Universitaria
Costa Rica, A.C.

The *REVISTA DE BIOLOGÍA TROPICAL* is a journal treating tropical biology that is published by the University of Costa Rica.

VIDA SILVESTRE NEOTROPICAL
World Wildlife Fund
1250 24th Street, N.W.
Washington, D.C. 20037 USA

VIDA SILVESTRE NEOTROPICAL is a biannual publication of the World Wildlife Fund with support from the W. Alton Jones Foundation, U.S. Fish and Wildlife Service, and the International Union for the Conservation of Nature and Natural Resources. It publishes high quality papers on wildlife research and management in the Neotropics.

ZOOLOGICA
New York Zoological Society (Publisher)
Bronx, New York 10460 USA

ZOOLOGICA is a scientific journal that was published at irregular intervals from 1907-1974. Its contributors were primarily scientists affiliated with the New York Zoological Society. Many of the articles resulted from research conducted in Trinidad, Venezuela, and Guyana.

MAPS

Map Link
25 East Mason
Santa Barbara, California 93101 USA
Telephone: (805)-965-4402

The Pacific Travellers Supply is the retail division of Map Link and advertises detailed maps and guide books for Latin America. Although their stock of maps varies according to availability, they usually have 1 : 250,000 scale maps or better for every South American country except Paraguay and the Guianas.

Map Link publishes *THE WORLD MAP DIRECTORY* which is updated annually. It lists maps for nearly every country in the world, including country, city, regional, topographic and specialty maps. Listings identify publisher, scale, and date of each map. The Directory also serves as a catalog.

South American Explorers Club
1510 York Street
Denver, Colorado 80206 USA
Telephone: (303)-320-0388

The South American Explorers Club puts out a catalog that features a wide selection of useful maps of Latin America.

Bradt Publications
41 Nortoft Road
Chalfont St. Peter
Bucks SL9 0LA
England

Bradt Publications publishes a catalogue titled *Travel Guides and Maps*, featuring many items pertinent to South and Central America.

Central Intelligence Agency
Public Affairs Office
Washington, D.C. 20505 USA
Telephone: (803)-351-2053

The U.S. Central Intelligence Agency (CIA) has a wide variety of excellent maps that can be purchased by the general public. To obtain a list of the map titles and prices, call or write and ask for the catalog *CIA Maps and Publications Released to the Public*.

World Wide Books And Maps
736 A Granville Street
Vancouver, British Columbia
Canada V6Z 1G3

World Wide Books And Maps specializes in maps and travel guides, and offers a wide selection related to Latin America.

ORGANIZATIONS

The American Forestry Association
P.O. Box 2000
Washington, D.C. 20013 USA
Telephone: (202)-667-3300

The American Forestry Association (AFA) is a citizen conservation organization that was founded in 1875 to foster the protection, wise management, and enjoyment of forest resources in America and throughout the world. The AFA attempts to maintain and improve the health and value of trees and forests, and to attract and encourage the interest of citizens, industry, and government. The AFA seeks to accomplish its goals through action-oriented programs, information and communication.

One of the largest, current projects of the AFA is Global ReLeaf. It is a national education, action, and policy campaign designed to improve the Earth's environment through more and better trees and forests. The AFA publishes a bimonthly magazine called *AMERICAN FORESTS*. The November/

December 1988 issue was devoted entirely to tropical deforestation.

Basic Foundation
P.O. Box 47012
St. Petersburg, FL 33743 USA
Telephone: (813)-526-9562

The Basic Foundation is a non-profit corporation that was founded in 1970 to augment efforts throughout the world at balancing population growth with natural resources. It gives financial support to a variety of activities aimed at educating the public about the threat of overpopulation. The immediate concern of the Basic Foundation is to save the tropical forests of the world. It is attempting to do this in a number of ways.

The Basic Foundation publishes and donates educational materials to schools. It also manufactures and distributes a variety of items (clothing, posters, notecards, etc.) bearing environmental and conservation messages. Proceeds from the sale of these items are used to further support educational projects. One of the most effective strategies the foundation uses to convey a conservation message and the importance of rainforests, is by organizing natural history tours into tropical forests at prices lower than the usual commercial rates. For example, advertised in the 1989 Rainforest catalog published by the Basic Foundation were nature trips to Monteverde (Costa Rica), Ecuador, and Venezuela.

The Basic Foundation is also working in conjunction with a Costa Rican conservation organization called Arbofilia. Their purpose is to establish and maintain tree nurseries on small farms in Costa Rica. The trees, including endangered native timber species and high quality fruit trees, are used to reforest degraded watersheds and areas not suitable for agriculture. Local farmers donate time and land and are trained in sylvicultural techniques.

Bat Conservation International
P.O. Box 162603
Austin, TX 78716 USA
Telephone: (512)-327-9721

Bat Conservation International (BCI) was established to educate people about the important role that bats play in the environment. BCI has been responsible for saving some of the world's most important bat caves and habitats. Although BCI was not established specifically for the conservation of tropical forests, the results of its efforts provide exactly that. Many bats are keystone species upon which thousands of other plants and animals rely for their own survival. Bats are responsible for up to 95% of the seed dispersal essential to tropical forest regeneration. If once-cleared rainforest is ever going to have a chance to grow back, the bat component of the fauna will have to remain intact. Additionally, bats pollinate some of the world's most valuable plants.

BCI is the world center for education, management, and research initiatives required for the conservation of bats. BCI serves as the international resource center for information about bats, and each year assists countries around the globe with regional education programs. Numerous research and conservation strategies are also planned and implemented as a direct result of BCI collaboration. The conservation of healthy bat populations is an integral part of preserving the diversity and quality of life on earth. BCI puts out a quarterly publication called BATS.

Better World Society
1100 Seventeenth Street, N.W.
Suite 502
Washington, D.C. 20036 USA
Telephone: (202)-331-3770

The Better World Society (BWS) is an international non-profit membership organization that attempts to increase individual awareness of global issues that are directly related to the sustainability of human and other life on Earth. Among the chief issues of the BWS are stewardship of the Earth's environment and efficient use of its resources. The BWS produces, commissions, acquires and distributes television programming that addresses issues of global concern. Such programming is sometimes supplemented by information kits, audio tapes, books, and other publications.

The BWS markets a number of video documentaries on global environmental issues. Two related to tropical deforestation and rainforests are *Jungle Pharmacy* and *Voice of the Amazon*. The first, highlights efforts to preserve native cultures and habitats that have developed lifesaving 'tropical pharmaceuticals' over thousands of years. The second, highlights the life and struggle of Brazilian conservationist Chico Mendes, who was murdered in late 1988. Chico Mendes was an environmental leader in Brazil and proponent of the use of sustainable 'extractive reserves' of rainforest as opposed to their destruction to produce pasture land.

Canyon Explorers Club
1223 Frances Avenue
Fullerton, CA 92631 USA

The Canyon Explorers Club (CEC) is a tax exempt, non-profit corporation that was started in 1972 and is based in California. Its purpose is to explore remote areas of the world. Members believe in the preservation of wilderness areas and respect the dignity of native cultures. Costs to members are minimized by booking group arrangements and planning trips well in advance. No member of the CEC may profit from any trip or outing.

The trips organized by the CEC vary greatly with regards to physical conditioning and endurance required of the participants. Some are no more strenuous than camping out, while others require a technical knowledge of mountain climbing. They range in duration from afternoon outings to month-long trips. Although the CEC runs trips throughout the world, during the 1989-90 period Latin American ventures included Guatemala, Belize, Venezuela, Peru, Costa Rica, and Brazil. Most trips originate from southern California, but with leader permission members can join a trip at the most convenient location.

The cost of becoming a member of the Canyon Explorers Club is $20, with annual dues of $10 thereafter (based on June 1989 Newsletter). A newsletter detailing upcoming trips, reporting on past trips, and featuring club activities in general is published three times a year (February, June, and October). Costs of specific trips are provided in the newsletters and vary according to location, activities, and duration. An example from the June 1989 Newsletter lists an 11 day trip to Costa Rica for $2,050.

Conservation International
1015 18th Street, N.W.
Suite 1002
Washington, D.C. 20036 USA
Telephone: (202)-429-5660

Conservation International (CI) is a non-profit conservation organization that was established in 1987. CI provides resources and expertise to private organizations, government agencies, and universities of Latin American and Caribbean countries in an effort to develop the capacity to preserve critical habitats. Through collaboration with local, national, and international organizations, CI works towards creating and managing ecosystem reserves. CI is headquartered in Washington, D.C., but maintains professional staff in Mexico, Costa Rica, Peru, Bolivia, and Brazil.

Conservation International publishes a quarterly newsletter called *TROPICUS*, which addresses topics concerning rainforest conservation. It also co-publishes the Orion Nature Quarterly magazine with the Myrin Institute.

Environmental Defense Fund
257 Park Avenue South
New York, New York 10010 USA
Telephone: (212)-505-2100

The Environmental Defense Fund (EDF) is a non-profit organization that was established more than 20 years ago. Its goals are to protect the earth's environment by providing lasting solutions to global environmental problems. By using the law combined with scientific research, the EDF attempts to find sustainable, non-destructive methods of addressing social and economic needs. The EDF works on a variety of environmental issues, some of which involve the conservation and wise use of tropical forests.

Regarding rainforests, the EDF was instrumental in the formation of a new department in the World Bank which promotes economically and environmentally sound loans. It also helped in the reauthorization of the Endangered Species Act, with the addition of some of its own measures. The EDF has also acted with Brazilian non-governmental organizations to establish an 'extractive reserve' program in Brazil. Such reserves are protected and then used for the continued harvesting of sustainable commodities such as Brazil nuts, rubber, and others.

The EDF puts out a quarterly newsletter and an annual report that can be obtained at the above address. They also can provide printed information and articles on rainforests, and on internationally-financed development in the third world.

Food And Agriculture Organization
Liaison Office For North America
1001 22nd Street, N.W.
Washington, D.C. 20437 USA
Telephone: (202)-653-2402

The Food And Agriculture Organization (FAO) is an organization of the United Nations. In 1985, it adopted a Tropical Forestry Action Plan with the goals of stemming the cosmopolitan destruction of tropical forests and implementing actions to preserve and efficiently use these natural resources. More specifically, the FAO is attempting to increase food production, improve methods of shifting cultivation, ensure sustained utilization of the forest, obtain better energy performance from fuelwood, and develop income and employment from associated sectors such as tourism and farm industries. Specific details of the FAO's Tropical Forest Action Plan an be obtained from the above address.

Global Tomorrow Coalition
1325 G Street, N.W.
Suite 915
Washington, D.C. 20005-3104 USA
Telephone: (202)-628-4016

The Global Tomorrow Coalition (GTC), established in 1981, is a national, non-profit alliance of organizations. Its members include private citizens and more than 110 U.S. organizations. The main function of the GTC is to educate teachers, policymakers, community groups, and concerned citizens about critical world problems affecting health and security. These global issues include environment, resources, sustainable development, and population.

RAINFOREST INFORMATION SOURCES

The GTC accomplishes its goals in some of the following ways. It publishes and distributes informative materials, produces teacher packets on such topics as Tropical Forests and Biological Diversity, briefs U.S. policymakers on global problems, holds teacher training workshops, promotes liaison between the U.S. and Third World non-governmental organizations, and works with U.N. agencies and international commissions.

The GTC publishes a newsletter called *INTERACTION*. This publication updates GTC members and legislative events that are related to the GTC's main areas of concern.

Greenpeace
Tropical Forests Campaign
1436 U Street, NW
No. 201-A
Washington, DC 20009 USA
Telephone: (202)-462-8817

Since its beginning, Greenpeace has been best known for its work on ocean ecology, toxic waste, and nuclear issues. Greenpeace recently expanded its agenda by launching hte Tropical Forests campaign. Greenpeace will use its strengths as an activist organization to identify and confront the major engines driving deforestation in the tropics. Greenpeace will focus attention on development, trade, and investment schemes which promote the destruction of tropical forests. Special attention will be directed to logging and timber trade industries as important catalysts of deforestation. In addition, Greenpeace is working with other organizations to support positive alternatives to the present destructivve models for 'development'.

The Greenpeace Tropical Forests Campaign will send you a fact sheet on Tropical Rainforests and information on how to join the Rainforest Activist Team if you send your name, address, and phone number. Greenpeace publishes the *GREENPEACE MAGAZINE* six times per year. The magazine is sent to supporters.

International Rivers Network
301 Broadway, Suite B
San Francisco, CA 94133 USA
Telephone: (415)-986-4694

The International Rivers Network (IRN) was established several years ago as a project of the Tides Foundation (see page 344). The purpose of IRN is to protect the major river systems of the world, and to encourage their wise management. It attempts to accomplish this through a global system of communications that provides for the exchange of information necessary to halt or prevent destructive river projects in particular areas. Information is gathered from and made accessible to environmentalists, engineers, economists, human rights groups, and tribal rights groups in 80 countries. The construction of hydroelectric dams on the Amazon River and its tributaries is one of the IRN's major concerns.

The IRN publishes a bimonthly newsletter called the *WORLD RIVERS REVIEW*. This presents the latest information and status of various river projects, along with additional related information of interest.

International Society For The Preservation Of The Tropical Rainforest
3302 N. Burton Avenue
Rosemead, CA 91770 USA
Telephone: (818)-572-7273

The International Society for the Preservation of the Tropical Rainforest (ISPTR) is a non-profit organization attempting to bring attention to the destruction of the world's tropical forests, and aid the animal inhabitants that suffer as a result of this destruction. The ISPTR has formed a division called the Preservation of the Amazon River Dolphin (PARD), which has attempted to relocate individual dolphins from endangered habitats. Through the use of educational displays and exhibitions in tropical countries, ISPTR/PARD attempts to educate nationals of the detrimental effects of the wild animal and skin trade while encouraging conservation tourism and the protection of natural habitats.

Another project the ISPTR/PARD is working on is a plan for Species Survival Sanctuaries in the Amazon Basin. These sanctuaries would be combination hospitals, museums, and research stations for animals that are stranded, injured, or diseased as a result of rainforest destruction, development, and river pollution.

Learning Alliance
494 Broadway
New York, NY 10012 USA
Telephone: (212)-226-7171

The Learning Alliance is a non-profit organization that was established in 1984. It attempts to make available to people, through its education programs, those resources necessary for making informed and intelligent decisions on today's issues. It is particularly concerned with various environmental issues including clean water and fresh air. Among the rainforest-related events that the Learning Alliance sponsors are slide shows of various tropical countries such as Costa Rica and Belize. It also sponsors educational travel to such countries.

National Audubon Society
International Program
801 Pennsylvania Ave., S.E.
Washington, D.C. 20003 USA
Telephone: (202)-547-9009

The National Audubon Society (NAS) is a non-profit conservation organization that was first started in the United States in 1905. It is supported by membership dues and contributions, and currently has more than 500,000 members. The conservation work of the NAS is carried out at local, state, regional, national, and international levels.

The broad goals of the NAS are the wise use of land and natural resources and the protection of wildlife and environmentally critical areas. This is accomplished in a number of ways including the management of a system of wildlife sanctuaries, education of the general public and decision-makers, influencing government agencies and legislators, and conducting scientific research on various bird species and their habitats. Several publications are put out by the NAS including *Audubon* magazine, *American Birds, the Audubon Activist, Action Alerts,* and the *Audubon Wildlife Report.* There is also a quarterly television production and a line of video software for educators.

Audubon has an active international program and has had international chapters for nearly 30 years. Audubon Societies exist today in Venezuela, Panama, Belize, Guatemala, Mexico, and Puerto Rico. Contact with these international chapters can often provide accurate, up to date information about conditions within a country, as well as suggestions regarding interesting localities that most tourists wouldn't know about. In some cases, as in Venezuela, the international Audubon office may also have a large selection of natural history books for sale.

Many U.S. Audubon chapters are focusing on specific international issues in their own conservation programs. Details are provided in a report titled *Audubon International Network* and can be obtained from the NAS's International Program Office (see above address). Another booklet available at the same office is *Foreign Assistance Action Project*. This project is aimed at developing an educated grassroots group, actively working and speaking out for "ecologically sustainable development". The booklet outlines how interested parties can go about accomplishing this goal. The NAS also puts out the *Tropical Forest Action Packet*, which supplies a list of educational resources available from many other organizations, as well as being extremely informative regarding rainforest issues.

Natural Resources Defense Council
1350 New York Avenue, N.W.
Washington, DC 20005 USA
Telephone: (202)-783-7800

The Natural Resources Defense Council (NRDC) is an environmental group with a staff of 80 lawyers, scientists, and environmental specialists, which maintain offices in New York, Washington, San Francisco, Los Angeles, and Honolulu. The NRDC provides legal representation and policy analysis for environmental causes and has greatly helped to shape the current environmental laws.

The NRDC has been strongly involved with the conservation of tropical forests in Hawaii, Puerto Rico, and the Virgin Islands. An affiliate organization, The Tropical Forests Working Group (TFWG), deals with tropical forest protection, legislation, and conservation on a more global scale. The TFWG publishes a newsletter that reports on current events with regards to tropical forests. It presents up to date information from a wide variety of sources. The TFWG can be contacted c/o the NRDC at the above address.

Nature Conservancy
International Program
1785 Massachusetts Avenue, N.W.
Washington, D.C. 20036 USA
Telephone: (202)-483-0231

The Nature Conservancy (NC) is a non-profit organization that was incorporated in 1951 for scientific and conservation purposes. The Conservancy maintains an international membership and is devoted to the global preservation of natural diversity. It locates, protects, and maintains the best examples of natural communities, ecosystems, and endangered species in the world. The Conservancy and its members have been responsible for the protection of over four million acres of habitat in the U.S. alone. It has assisted with the protection of many times that amount in Canada, Latin America, and the Caribbean. It presently maintains the world's largest privately-owned nature preserve system, with more than 1,100 preserves. The NC's International Program established in 1981, aids and collaborates with conservationists in other countries, increasing their capacity to identify and protect critical natural areas.

The Nature Conservancy's International Program has concentrated on Latin America and the Caribbean. It has established a network of computerized conservation data centers (CDCs), which by 1990 were located in Bolivia, Brazil, Colombia, Costa Rica, Netherland Antilles, Panama, Paraguay, Puerto Rico, Peru, and Venezuela. Through these CDCs, and in collaboratiuon with various Latin American partner organizations, the NC helps scientists in determining which species and habitats need protection and how others can best be used. The Latin American Program works with private, non-profit national and regional organizations, with government agencies, and with other international institutions throughout Latin America to conserve biological diversity.

The Nature Conservancy publishes *THE NATURE CONSERVANCY MAGAZINE* and assorted country fact sheets and other publications describing its work and conservation projects within and outside of the United States.

Partners of the Americas
1424 K Street, N.W. #700
Washington, D.C. 20005 USA
Telephone: (202)-628-3300
Telex: 64261

The Partners of the Americas organization is a private, nonprofit, voluntary group with the goals of enhancing social and economic development in the Western Hemisphere. Partners of the Americas originated as the Partners of the Allicance program that was established in 1964 within the Agency for International Development (USAID). Partners of the Americas became fully private in 1970, although they continued to receive funding from USAID. Today, the Partners of the Americas has greater than 20,000 volunteer members spread throughout the United States, Latin America, and the Caribbean that conduct projects with an estimated value of close to $80 million a year.

Partners of the Americas functions by pairing volunteers from states of the United States with volunteers in areas throughout Latin America and the Caribbean. Both volunteer groups then work in conjunction to accomplish various projects which are aimed at improving the standard of living in some way in the host country. While these programs are usually aimed at health, education, agriculture, it has provided the Partners of the Americas organization with an extensive knowledge of people and localities that are 'off the beaten track' in Latin America.

One of the most useful projects conducted by the Partners of the Americas in conjunction with the Tinker Foundation was the formation of a directory to conservation and related organizations in Latin America and the Caribbean. This book, *A DIRECTORY OF NATURAL RESOURCE MANAGEMENT ORGANIZATIONS IN LATIN AMERICA AND THE CARIBBEAN* (see page 243), covers most of the countries of Central and South America and provides a wealth of valuable information to anyone interested in conservation or natural resource management in Latin America.

Peace Corps
1990 K Street, N.W.
Washington, DC 20526 USA
Telephone: (800)-424-8580

The Peace Corps is an international volunteer organization that has offered natural resource assistance to developing countries over the past 25 years. All of the Peace Corps' natural resource programs are designed in response to host country requests for assistance. The Peace Corps is currently conducting projects in the following areas related to natural resources: Environmental Education, National Parks/Biological Diversity, Wildlife Management, Agroforestry, Forestry Extension, and Forestry Management. Among the Latin American countries with Peace Corps programs in natural resources are: Belize, Costa Rica, Ecuador, Guatemala, Honduras, and Paraguay. The Peace Corps volunteers and staff can provide in-depth, specific information about various localities where projects are being carried out, or where they have been completed in the past.

Rainforest Action Network
301 Broadway
Suite A
San Francisco, CA 94133 USA
Telephone: (415)-398-4404

Since it was founded in 1985, the Rainforest Action Network (RAN) has been working to protect tropical rainforests and the human rights of those living in and around those forests. From the beginning, the Network has played a key role in strengthening the worldwide rainforest conservation movement through supporting activists in tropical countries as well as organizing and mobilizing consumers and community action groups throughout the United States. Their first challenge was to bring the plight of the rainforests to public attention through education, communication, and direct action. They began by convening the first international rainforest conference where activists from 35 organizations formulated a plan of action. This conference was followed by others which have helped to catalyze the growing world rainforest movement. Through media campaigns, conferences, and publications, the RAN's efforts have helped to make the rainforest issue the *cause celebre* that it has become in the U.S.

Within North America alone, there are now over 120 Rainforest Action Groups (RAGs) associated with RAN. RAGs carry on the essential grassroots work of educating local communities and gathering the critical mass needed to exert effective pressure for change when and where it is needed. RAN's monthly *ACTION ALERT* and quarterly *WORLD RAINFOREST REPORT* keep the RAGs and RAN's individual members informed about the assaults on the rainforests and what we as individuals can do about them.

Consumer education continues to be an essential activity of the Network. RAN's current focus in this regard is the tropical timber trade. Their research into the institutions and industries responsible for rainforest destruction enables a growing number of people to turn concern into effective action.

RAN works with environmental and human rights groups in 60 countries, sharing information and coordinating the U.S. sector's role in worldwide campaigns to protect the rainforests and their inhabitants. Through financial support and networking services, RAN supports the efforts of indigenous and environmental groups in tropical countries to achieve ecologically sustainable solutions within their own regions. RAN is distinguised by its emphasis on a grassroots approach, its networking capabilities, and its commitment to mobilizing citizen activists to respond quickly and directly to the forces that threaten the rainforests and all of us.

Rainforest Alliance
270 Lafayette Street
Suite 512
New York, N.Y. 10012 USA
Telephone: (212)-941-1900

The Rainforest Alliance (RA) is a non-profit educational organization that was founded in 1986. Its main purpose is to increase public awareness of both the role that the United States plays in the fate of tropical forests, and how the destruction or conservation of rainforests will in turn affects us. The RA also works increasingly with the public and private sectors to promote policies and actions that encourage forest conservation.

Besides its public outreach program, some of the RA's other projects include: The Periwinkle Project, which is building a constituency of health-related professionals to increase research on the medicinal uses of tropical plants; The Tropical Timber Project, which is examining the tropical woods imported into the U.S., and promoting sustainable forestry practices; The Kleinhans Agroforestry Fellowship, for research on sustainable uses of forestry products; and support for AMETRA 2001 (Aplicacion de Medicina Tradicional), a health cooperative run by and for forest dwellers in the Peruvian Amazon that uses traditional knowledge of forest plants.

The RA publishes a periodic newsletter called *THE CANOPY*, which reports on many rainforest-related topics, including socioeconomic and political issues. *HOT TOPICS FROM THE TROPICS"*, a listing of events and other happenings, is published bi-monthly.

Rainforest Health Alliance
Fort Mason, Building E
San Francisco, CA 94123 USA
Telephone: (415)-921-1203

The Rainforest Health Alliance (RHA) is a non-profit organization dedicated to saving the earth's tropical rainforests in an effort to protect and maintain their tremendous health benefits for humankind. Organized under the auspices of Earth Island Institute, RHA is action-based and project-oriented. Their current projects are focused in three educational areas: rainforest expeditions, children's education materials, and developing support from within the natural products business community and the pharmaceutical industry. The RHA is currently building a resource base of information about native and medicinal plants and the indigenous peoples of rainforests, and is developing fact sheets on medicinal plants and their traditional origins.

RARE Center For Tropical Bird Conservation
19th and the Parkway
Philadelphia, PA 19103 USA
Telephone: (215)-299-1182

The RARE Center For Tropical Bird Conservation (hereafter to be referred to as the RARE Center) is a non-profit organization affiliated with the Academy of Natural Sciences in Philadelphia. Its aims are to prevent the loss of endangered tropical bird species and their habitats. Founded as RARE in 1973, the organization's original goals were to protect endangered species in general. RARE then worked to aid in the preservation of a variety of animals including primates, whales, snow leopards, jaguars, sea turtles, and endangered birds. During the 1980s RARE worked to preserve endangered ecosystems including coral reefs, cloud forests, and rainforests. It also developed environmental education programs for the Caribbean and Latin America. RARE changed its name to the above in 1986 and also focused its efforts to endangered bird conservation in the tropics.

The RARE Center actively works on a small number of projects. It attempts to make lasting conservation contributions by combining publicity/education campaigns and basic research with habitat protection. The RARE Center is currently concentrating on species of cracids, neotropical parrots, and both tropical wetland and highland birds. It is involved in land acquisition at the Monteverde Cloud Forest Reserve in Costa Rica (see page 171) and a study of the seasonal altitudinal distribution of the resplendent quetzal there. It has conservation programs centering around endangered parrots in St. Vincent and Dominica, the white-winged guan in Peru, and the el oro parakeet in Ecuador. In addition, the RARE Center has also actively supported surveys of birds in Ecuador and the production of the book *BIRDS OF SOUTH AMERICA*.

Sierra Club
730 Polk Street
San Francisco, California 94109 USA
Telephone: (415)-776-2211

The Sierra Club is a non-profit conservation organization that was founded by John Muir in 1892. It is currently one of the largest and most influential environmental organizations in the world. By intensive lobbying, the Sierra Club attempts to influence decisions made by various levels of government that affect the land and natural resources of the United States and the world. The Sierra Club's efforts have been primarily devoted to the United States, although they maintain contact with conservation organizations throughout the world. Various literature regarding tropical forest destruction is available from the Sierra Club.

Smithsonian Institution
National Museum of Natural History
Dr. Rob Bierregaard/Victor Bullen
NHB-106
Washington, DC 20560 USA
Telephone: (202)-786-2821

The National Museum of Natural History of the Smithsonian institution collects, preserves, studies and displays specimens from the natural world and objects made by its inhabitants. There are over 80 million objects in the museum's inventory. Some 120 scientists conduct research in various fields of natural history and paleontology throughout the world. The Museum's Biological Diversity in Latin America (BIOLAT) Program has field sites in Brazil, Peru, and Bolivia.

The administration of the Biological Dynamics of Forest Fragments Project, formerly called the Minimum Critical Size of Ecosystems Project, was transferred from the World Wildlife Fund (WWF) to the Smithsonian Institution in July 1989. The Forest Fragments Project is a binational study between the Smithsonian Institution and the Institute Nacional de Pesquisas da Amazonia (INPA) of Brazil. Now in its 11th year, the goal of the project is to determine the effect of cutting the forest into various sized fragments. More details are available on page 325.

Scientific articles from the Forest Fragments Project's technical series, annual reports, popular articles, as well as information describing project protocol, organization, field sites, living conditions and other aspects, can be obtained from Dr. Rob Bierregaard-Principal Investigator, and Victor Bullen-Washington Coordinator, at the above address.

South American Explorers Club
1510 York Street
Denver, CO 80206 USA
Telephone: (303)-320-0388

South American Explorers Club
Casilla 3714
Lima 100
Peru
Telephone: 31-44-80

South American Explorers Club
Apartado 21-431
Eloy Alfaro
Quito, Ecuador
Telephone: 56-60-76

South American Explorers Club (continued)

Street
Address: Avenida Republica de
Portugal 146
Breña, Lima

Street
Address: Toledo 1254
La Floresta, Quito

The South American Explorers Club (SAEC) is an extremely useful organization to tourists, biologists, backpackers, or anyone who is interested in Latin America. The SAEC was founded in 1977 in Lima, Peru, where one of its South American offices is located today. The SAEC is a non-profit, non-political, non-sectarian organization that has a number of related and worthwhile goals. The one that perhaps best exemplifies its purpose is "To further the exchange of information among scientists, adventurers, and travelers of all nations with the purpose of encouraging exploration throughout the continent of South America". Other interests of the SAEC are promoting wilderness conservation and wildlife protection, supporting and documenting field and adventure sports, aiding scientific, educational, or cultural programs, and making available reliable information on organizations in South America which offer services to travelers, scientists, and outdoorsmen.

The SAEC accomplishes its goals in several ways. It publishes a quarterly educational and scientific magazine called the *SOUTH AMERICAN EXPLORER*. A subscription is included in the club's regular membership fee. The *SOUTH AMERICAN EXPLORER* features articles on scientific studies, adventures, sports, historical explorations, biographies of South American explorers, and descriptions of unusual places. It also publishes letters, tips, and notes to travelers, classified ads, a catalog of books and tapes, and trip reports.

The trip reports are one of the most useful features of the club. They are typically 1-3 page reports written by travelers from all walks of life about the trips they have taken to various Latin American locations. The reports describe transportation, accomodations, costs, contacts made, food eaten, wildlife observed, etc. They can be extremely helpful, depending on the conscientiousness of the writer, but in almost all cases are informative. Many trip reports are available for locations in Peru, with Spanish-speaking South American countries the next most common category. However, there are also some reports about areas in Brazil and Central America. Trip reports are made available to SAEC members for a very nominal photocopying and postage charge.

Members of the SAEC are also entitled to services at the Lima clubhouse located at 146 Avenida Portugal in Breña (Lima, Peru), and at the Quito clubhouse located at Toledo 1254 Las Floresta (Quito, Ecuador). Among these services are access to information sources including maps, trip reports, a library with both English and Spanish books, and contacts of knowledgeable people available for specialized information. The clubhouse staff will help members reconfirm flights, make hotel reservations, advise and provide logistical support for expeditions, and provide storage for equipment and possessions.

The annual membership fee is $25. This entitles members to four issues of the *SOUTH AMERICAN EXPLORER*, a discount on items sold by the SAEC, and Lima clubhouse priviledges. The information and services available through the SAEC are well worth the price, especially if one is hesitant about travelling to Latin America and without contacts. The SAEC will also attempt to match members with traveling companions, if requested.

Summer Institute of Linguistics
7500 West Camp Wisdom Road
Dallas, TX 75236 USA

The Summer Institute of Linguistics is a religious organization with facilities throughout the world. Also known as the Wyckoff Bible Translators, they sometimes can be helpful in providing information and limited logistical support in various areas where they have field facilities. For many years they maintained a residence at Limoncocha, Ecuador, which has since closed. According to recent information, they maintain field outposts near Pucallpa, Peru and in Brazil.

Tropical Forests Forever
P.O. Box 69583
Portland, OR 97201 USA
Telephone: (503)-227-4127

Tropical Forests Forever (TFF) is a non-profit organization that was established in 1988. Its goal is conserving tropical forests for future generations. It sponsors monthly program meetings that provide in-depth information for hundreds of activists working to conserve tropical forests. It has presented a wide range of lecture and slide programs including experts on Latin American affairs, forestry in the Caribbean, and tropical wildlife. Members have worked to inform the public about the importance of tropical forests.

The TFF publishes a bimonthly newsletter called *TROPICAL FOREST NEWS*. This newsletter informs readers about current legislation, country profiles, tropical events schedule, amazing jungle facts, and others.

United Nations Information Centre
1889 F Street, N.W.
Ground Floor
Washington, DC 20006
Telephone: (202)-289-8670

The United Nations Information Centre can provide a wealth of information and publications regarding various U.N. programs being conducted by agencies and organizations within the U.N. System. Examples of such programs, agencies, and organizations are: the United Nations Environment Programme (UNEP), the Food and Agriculture Organization (FAO), the World Health Organization (WHO), and the United Nations Educational, Scientific and Cultural Organization (UNESCO). The UNEP Environment Brief No. 3 specifically treats the disappearance of tropical forests.

Wildlife Conservation International
New York Zoological Society
Bronx, N.Y. 10460 USA
Telephone: (212)-220-5090

Wildlife Conservation International (WCI) is a division of the New York Zoological Society (NYZS). Formerly called the Animal Research and Conservation Center, WCI serves as the international conservation program of the NYZS. Through the support of a full-time staff of field biologists, WCI attempts to increase the understanding of endangered species and their ecosystems. The knowledge obtained through their programs and scientists is then applied to conservation. WCI is a privately funded, tax-exempt organization, with all funds raised going to support field research and conservation.

Wildlife Conservation International is truly international in scope with support currently given to 82 projects in over 37 countries in 5 continents. The WCI scientists work primarily in the third world and are taking part in a variety of projects in Latin America. Many of these projects involve flora and/or fauna found in the neotropical forests. In 1988 WCI supported programs in Belize, Venezuela, Colombia, Peru, Ecuador, Paraguay, Chile, and Argentina. WCI maintains some of the most reknown scientists in their field. Their efforts have helped establish 50 national parks and nature preserves in 20 countries.

World Forestry Center
4033 S.W. Canyon Road
Portland, OR 97221 USA
Telephone: (503)-228-1367

The World Forestry Center (WFC) is a non-profit education organization. Its purpose is to increase understanding of the importance of well-managed forests and their related resources. The WFC demonstrates and interprets the benefits of conserving the forest environment through its publications, educational programs, exhibits, and its architecture. The WFC is also dedicated to the conservation of trees, soil, water, wildlife, and other natural resources. This is accomplished through scientific research, demonstrations, and the distribution of forestry information.

The World Forestry Institute (WFI) is a recently established branch of the WFC. The WFI serves as an international forestry and forest products clearinghouse. WFI identifies, collects, interprets and disseminates information on domestic and international forestry research and development. Much of this is accomplished through the Institute's Forest Information Network and Directory (FIND), which accesses both major databases as well as lesser-known electronic and non-electronic international data sources. WFI's ongoing activities include conferences, symposia, and exhibits on international forestry and natural resource issues.

The WFC publishes a quarterly magazine called *FOREST WORLD*. It is put out for the education and enjoyment of WFC members who have an interest in the conservation of the world's forests. *FOREST WORLD* treats topics pertaining to domestic and international forests and conservation.

World Resources Institute
1709 New York Avenue, N.W.
Washington, D.C. 20006 USA
Telephone: (202)-638-6300

The World Resources Institute (WRI) is an independent, non-profit, non-partisan organization that functions as a policy research center. Their primary goal is to establish policies that preserve natural resources while continuing to encourage economic growth and to meet the needs of society. They work with both governments and the private sector, as well as environmental leaders and others. The WRI maintains a large interdisciplinary staff, in addition to collaborators in 30 countries.

The World Resources Institute is currently concerned with two major groups of problems: 1) The destructive effects of poor natural resource management on economic development and on the alleviation of poverty in developing countries; and 2) the recently evolved environmental and resource problems of global importance that threaten the economic and environmental interests of the U.S. and many other countries. WRI is conducting a number of policy research projects within the above areas of convern, including: a Program in Forests, Biodiversity, And Sustainable Agriculture, and a Program In Resource And Environmental Information. WRI also has a Center for International Development and Environment which provides services for developing countries in the sustainable management of natural resources.

WORLD RESOURCES is a biennial report on international conditions and trends in population, resources, and environment, published by WRI. A wide variety of other publications are also put out by WRI covering the themes of resources and development, energy and climate, and agriculture. A number of these titles are directly related to tropical forests (see pages 271-277 in Chapter 2 under 'Environmental, Political and Miscellaneous Titles' section.

World Wildlife Fund
1250 Twenty-Fourth Street, N.W.
Washington, D.C. 20037 USA
Telephone: (202)-293-4800
Telex: 64505 PANDA

The World Wildlife Fund (WWF) is a non-profit international conservation organization that was founded in 1961. Its main goal is to protect endangered wildlife and their habitats throughout the world. The WWF has national affiliates in over 23 countries and has supported more than 4,000 projects in 130 countries, including 260 national parks. The WWF accomplishes its conservation goals by several methods including the training of local scientists, supporting education of conservation topics, and by directly buying or obtaining land.

During the 1980s, the WWF targeted more than half of its program expenditures on conservation in Latin America. It supported over 500 projects advancing wildlife and habitat preservation in the region. Included among these programs are the establishment, maintenance, and protection of Manu National Park (Peru), whose 6,000 sq miles shelter nearly 10% of all bird species. The reintroduction into Brazil of captive-bred golden lion tamarins, an endangered primate. Through the Resource Management Education Program the WWF has developed materials and trained educators to teach basic environmental and ecological concepts to young students. In partnership with Brazil's Indian Protection Agency, WWF is studying traditional Kayapo Indian agricultural practices, which use sophisticated land conservation skills that may be valuable elsewhere in the tropics.

The World Wildlife Fund maintains an international trade-monitoring program called TRAFFIC. Its function is to curb and influence the illegal wildlife trade. The WWF also publishes a bimonthly newsletter called *FOCUS*, describing current WWF projects and recent events in conservation. The WWF puts out a Publication/Audiovisual Catalog listing many interesting and educational items. Some of the books listed in the 1988 catalog were *LATIN AMERICAN WILDLIFE TRADE LAWS, CROCODILE SKIN TRADE IN SOUTH AMERICA, THE INTERNATIONAL PRIMATE TRADE, PRIMATES AND THE TROPICAL FORESTS,* and *WORLD CONSERVATION STRATEGY.*

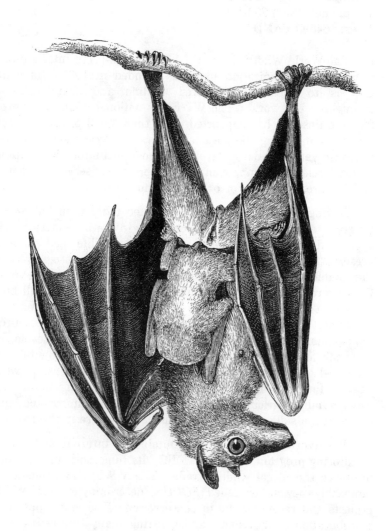

CHAPTER 3 'HANDS-ON' ORGANIZATIONS

INTRODUCTION

There is a completely different order of magnitude in appreciation of the rainforest community that is gained from actually visiting the jungle, as opposed to reading about it or watching it on a nature program on television. This appreciation jumps to a still higher level when one works in close contact with a scientist or researcher who is intimately acquainted with specific aspects of rainforest biology and ecology.

During the past 25 years several organizations have evolved that provide laymen and/or students with the opportunity to work side by side with scientists in the field. Although some of these organizations support a wide array of ecological, archaeological, and anthropological projects, the number of studies involving tropical forests are increasing. Presented below is a list of certain 'hands-on' organizations and a description of how they operate.

The American Ornithologists' Union
c/o Ornithological Societies of North America
P.O. Box 21618
Columbus, OH 43221 USA

The American Ornithologists' Union (AOU) publishes a bimonthly newsletter called the *ORNITHOLOGICAL NEWSLETTER*, which is received by members of several ornithological societies. There is a section of this newsletter called Requests For Assistance and Positions And Opportunities Available. In these columns, descriptive requests are listed for volunteers or assistants to work primarily on ornithologically related projects and research. Often there are opportunities for field work and sometimes these include Central and South America.

Caribbean Conservation Corporation
P.O. Box 2866
Gainesville, FL 32602 USA
Telephone: (904)-373-6441
Telex: 387530 CCC TAS

or

Caribbean Conservation Corporation
Apartado Postal 6975-1000
San Jose
Costa Rica
Telephone: 33-80-69
Telex: 3015 CESTA CR

The Caribbean Conservation Corporation (CCC) is an international public foundation that was formed in 1954. Its purpose is to support long-term studies and conservation of the Atlantic Green Turtle. Research of the CCC was directed by the famous zoologist and author Archie Carr, until his death.

The CCC has maintained its Green Turtle Research Station since 1954 at Tortuguero, on the Caribbean coast of Costa Rica. It was built on a black beach used as a breeding ground by the green sea turtles and now facilitates researchers during the seasonal monitoring of the western Caribbean green turtle colony.

The CCC uses 6-8 students and graduate students each year to assist in monitoring and tagging the sea turtles. In return for volunteering their time for the three-month turtle breeding season (approximately July 15th to September 15th), the students are provided room and board. Expected to secure their own transportation to Tortuguero, the students work typically at night, seven nights a week; they can then sleep late and have most afternoons free.

The CCC also accepts volunteers for 1-2 week periods from the general public in a program that is currently administered through the Massachusetts Audubon Society. Details on costs and how to participate can be obtained from the Massachusetts Audubon Society at the following address:

Massachusetts Audubon Society
South Great Road
Lincoln, MA 01773 USA
Telephone: (617)-259-9500

Earthwatch
680 Mt. Auburn Street
P.O. Box 403
Watertown, Massachusetts 02272 USA
Telephone: (617) 926-8200
Telex: 5106006452

Earthwatch is a non-profit, tax-exempt organization that was founded in 1971. Its purpose is to aid scientists and qualified investigators in their efforts to learn more about both the biotic and abiotic factors of the world we live in. Past projects supported by Earthwatch have been in the disciplines of anthropology, archaeology, astronomy, biology, botany, entomology, geology, and zoology, to name a few. Earthwatch obtains the majority of the finances needed to support such projects by a novel means they have termed 'participant funding'.

In participant funding, volunteers are solicited to help finance and form the labor force necessary for a given study. An expedition contribution is calculated according to the project budget and is paid by each participant. This contribution is a pro-rated share of the total expedition costs. These funds are used for field grants awarded to investigators, and to pay the costs associated with running the Earthwatch organization. In addition to the expedition contribution, volunteers must also secure their own transportation to the staging area of the project. The staging area is usually the closest city containing an airport to where the actual field work is to be conducted. Transportation costs will therefore vary according to the residence of the volunteer and the location of the study.

Earthwatch volunteers are not normally expected to have prior expertise in the discipline of a given project. It is the responsibility of the principal investigator to instruct, train, and use his volunteer work force effectively. Volunteers are grouped into teams, which are expected to provide their services in the field during previously selected dates. The length of the field session for each team usually averages 10-14 days. In the 1990s, Earthwatch foresees an increasing number of projects focusing on the problems of global environment.

Earthwatch publishes a bimonthly magazine that provides expedition descriptions, costs, and field dates for the upcoming year. In the 1989 season, expedition costs ranged from $800 to

$2,025. Project descriptions indicate the degree of physical difficulty expected and inform potential volunteers about the field conditions. Volunteers are sent an expedition briefing, which discusses the research and logistics in detail, after formally applying.

Since Earthwatch was founded, it has supported more than 716 projects in 89 countries all over the world. It has brought together more than 22,000 volunteers to investigators and has provided more than $12 million dollars in equipment and funds. Earthwatch is currently the third largest private source of funding in the United States for field research. In addition to supporting scientists and researchers, Earthwatch also provides fellowships and scholarships to teachers and students. This will be discussed in the next chapter.

Foundation For Field Research
P.O. Box 2010
Alpine, CA 92001-0020 USA
Telephone: (619)-445-9264

The Foundation for Field Research (FFR) was founded in 1982 as a non-profit organization whose goal is to bring together the needs of researchers with support from the public. The FFR selects and develops field reseaqch projects that are conducted as cooperative expeditions assisted by non-specialist volunteers. The FFR supports research in the following and other fields: anthropology, prehistoric and historic archaeology, art history, astronomy, botany, ecology, entomology, ethnobotany, ethology, folklore, folk medicine, geography, geology, herpetology, historic architecture, marine archaeology, marine biology, ornithology, paleontology, primatology, and wildlife biology.

Grants to researchers, as well as a portion of the FFR's overhead, are obtained through fees; or share-of-cost contributions are calculated for each project and paid by the volunteers. These contributions are tax deductible. The FFR publishes a periodical called the *EXPLORER NEWS*, which lists and describes current FFR projects in need of volunteers. In Vol. 4, Nos. 1 & 2, 1988, the necessary project contributions ranged from $95 for a two day weekend conducting archaeological research in California, to $1,285 for studying chimpanzees in Africa. The majority of studies supported the use of several teams of from five to twenty volunteers for a period of five to fourteen days each. Volunteers are expected to finance their own way to the stated assembly point.

International Research Expeditions
140 University Drive
Menlo Park, California 94025 USA
Telephone: (415) 323-4228
Telex: 5106002876 IREXP USA

International Research Expeditions (IRE) is a non-profit organization that brings together field research scientists with members of the public who wish to assist them. Volunteers are responsible for reaching a staging area and pay a contribution selected according to the project's budget. All contributions and project-related travel costs are tax deductible.

Research groups usually consist of 5-10 people, while volunteer terms may range from 1-4 weeks. The emphasis of IRE projects is to help preserve endangered species, increase our knowledge of the natural history of all organisms, and to better understand the earth's natural forces.

The 1990-91 IRE list of expeditions features 45 projects of various disciplines all over the world. Required contributions ranged from $200 for a one week session in Ohio to study pond ecology to $3,500 for a 21 day expedition to Ecuador studying the endangered Black Caiman.

Monteverde Institute
Tropical Biology Program
    administered by
Council on International Educational Exchange
205 East 42nd Street
New York, N.Y. 10017 USA

The Monteverde Institute is a non-profit educational organization that offers a summer academic program administered by the Council on International Educational Exchange (CIEE). The 1989 program was conducted from June 10th to August 9th and included courses in tropical biology, Spanish language instruction, and independent studies. Participating students are based at the field station at the edge of the Monteverde Cloud Forest Reserve in Costa Rica (see pages 166-171).

All courses are taught in English by doctoral level biologists who have conducted research at Monteverde. The tropical biology course introduces students to climate, geology, soils, biogeography, population biology, in addition to evolutionary patterns and interactions among plants and animals. Tropical biology labs and field work are conducted in various habitats and ecological life zones. Field trips include camping at one or more of Costa Rica's National Parks, as well as other sites of biological interest.

To be eligible for this program, students must have 3-4 semesters of college level biology courses or the equivalent, with a 2.75 grade average. Previous Spanish language instruction is also preferred. Credit for the student's program of study is determined by his home institution. The tuition for the 1989 program was $2,950, not including airfare. In-country transportation is provided by charter bus.

Organization For Tropical Studies (OTS)

North American Office
Box DM
Duke Station
Durham, North Carolina 27706 USA
Telephone: (919) 684-5774

Costa Rican Office
Apartado 16
Universidad de Costa Rica
San José
Costa Rica
Telephone: 36-66-96

The Organization For Tropical Studies, Inc. (OTS) is a non-profit consortium that was founded in 1963. The University of Costa Rica and seven U.S. universities formed the charter members; today there are 49 member institutions of higher learning. The purpose of OTS is to promote the wise use of natural resources in the tropics, and to provide leadership in education and research. OTS maintains three field stations with research facilities at diverse tropical forest sites in Costa Rica (see pages 171-178, and page 323). These facilities are available to researchers from both OTS institutions and nonmember universities for a nominal fee.

OTS instructs graduate students in the basics of tropical biology and ecology by offering an intensive field-oriented course in Costa Rica twice a year. Entitled Tropical Biology: An Ecological Approach, this eight-week course is taught by 10-15 instructors who are often course alumni. All graduate students are eligible to enroll in Tropical Biology, however only 20 spaces are available each term and acceptance is on a competitive basis. A course fee of $1000 is currently charged to students from OTS member institutions, while $2500 is charged to non-member institution students. This fee does not include travel costs to the staging area of San Jose, Costa Rica. Eight (8) semester hours are earned for the successful completion of the course with a transcript issued by the Universidad de Costa Rica.

The course begins with introductory lectures and orientation in San Jose. Following this, students are transported to six or more diverse sites, including the OTS field stations of the Wilson Botanical Garden, Las Cruces (wet mid-montane forest), Palo Verde (deciduous dry forest), and La Selva (lowland rainforest). Other potential sites include the Monteverde Cloud Forest Reserve, the agricultural experiment station and training center (CATIE) at Turrialba, and the national parks of Tortuguero (Caribbean beaches), Santa Rosa (deciduous dry forest) and Corcovado (lowland rainforest). At each location students receive an orientation of the area, followed by several days of group-oriented field problems. After becoming familiar with an area, course members are encouraged to conduct independent research projects.

The OTS Tropical Biology course is an intensive and demanding experience. Daily activities often begin early in the morning, ending with seminars in late evening. Research and training is carried out seven days a week, taking full advantage of time in the field. Participating students must have at least of four graduate level courses in biology, or approval from the course coordinator. Post-course research awards and tropical fellowships are offered on a competitive basis through OTS (see Chapter 4).

In the past several years, OTS has expanded its curriculum to include other field courses. Those scheduled on a regular basis include the following:

Tropical Biology: An Ecological Approach
Ecologia de Poblaciones
Tropical Agricultural Ecology

AgroForestry: Management Alternatives for Tropical Ecosystems
Principios Ecologicos para la Toma de Decisiones y el Manejo de los Recursos Naturales en America Latina
Agroecologia
Interdependence: Economic Development and Environmental Concerns in Tropical Countries
Ecologia de Poblaciones de Vida Silvestre

The language in which the course title appears is the language of instruction.

School For Field Studies
16 Broadway
Beverly, MA 01915 USA
Telephone: (508)-927-7777

The School for Field Studies (SFS) was established in 1980 as an educational institution. It offers its students a combination of formal instruction and actual field participation on environmental and biological research projects. All SFS courses are accredited and involve lectures, assigned readings, reviews of current literature, presentations, independent projects, and exams. Participants also learn about research methodology, teamwork, outdoor living, and the particular branch of science the project falls under.

The SFS courses are of one month or one semester (13 weeks) duration. The course team sizes range from 14 students/2 faculty on the one month courses to 24 students/4 faculty on the semester programs. Past participants have been students from private and public high schools, colleges, and universities, ranging in age from 16 to 27. Applicants for the one month courses must have completed their junior year of high school. Applicants for the semester courses should be at least 18 and have completed one or more college level courses in biology or ecology. To participate on a SFS introductory course, no science background or previous field training is needed. Financial aid is available based on need (see Chapter 4).

Semester courses are facilitated by the SFS research centers maintained in Australia, Kenya, and the Virgin Islands. Research is conducted throughout the year at these institutions, with the Center for Rainforest Studies (Queensland, Australia) specifically devoted to biological investigations of the rainforest. The SFS also has mobile bases located in the tropical forests of Costa Rica, Ecuador, and Mexico. One month summer courses on tropical ecology, ethnobotany, and experimental social ecology have been held at these sites, respectively, in 1989.

Smithsonian Institution
National Museum of Natural History
Rob Bierregaard/Victor Bullen
NHB-106
Washington, DC 20560 USA
Telephone: (202)-786-2821

Roger Hutchings, Director de Campo
Convenio INPA/WWF
Ecologia INPA
C.P. 478
69011 Manaus, AM; Brazil
Telephone: (092)-236-5568/8155

Volunteer internships of six months duration are sometimes available for upper-level college students or recent graduates who plan to go on to graduate programs in the biological sciences. The internship is carried out in Brazil as part of the Biological Dynamics of Forest Fragments Project.

The administration of the Biological Dynamics of Forest Fragments Project, formerly called the Minimum Critical Size of Ecosystems Project, was transferred from the World Wildlife Fund (WWF) to the Smithsonian Institution in July 1989. The Forest Fragments Project is a binational study between the Smithsonian and the Instituto Nacional de Pesquisas da Amazonia (INPA) of Brazil. Now in its 11th year, the goal of the project is to determine the effect of cutting the forest into various sized fragments. Twenty-three reserves, ranging in incremental sizes with replicates from 1, 10, 100 to 1000 hectares, and a control area that will remain in unbroken forest are being studied. Ten of the 23 reserves have been isolated by cattle pasture.

The effect of isolating the reserves is measured by surveying flora and fauna before, during, and after the forest clearing, and by comparing the survey data of the fragments to those of the continuous forest in the control area. The results of this study will provide much needed data on how large an area must be set aside and protected to establish a protected area that will maintain the majority of its biota. Results will also indicate which species are sensitive to isolation and how they might be managed in protected areas that are too small.

The project's extensive infrastructure also facilitates research in Amazon natural history. Seven field camps are located near the project reserves about 80km north of Manaus, Brazil. The reserves have a lengthy trail system. Investigators interested in conducting research on forest fragmentation or Amazonian natural history should contact Dr. Bierregaard at the above address.

Scientific articles from the Forest Fragments Project's technical series, annual reports, popular articles, as well as information describing project protocol, organization, field sites, living conditions and other aspects, can be obtained from Dr. Richard O. Bierregaard, Principal Investigator, and Victor Bullen, Washington Coordinator, at the above address.

Smithsonian Research Expeditions
Smithsonian Associates Research Expedition Program
Smithsonian Institution
490 L'Enfant Plaza, S.W.
Suite 4210
Washington, D.C. 20024 USA
Telephone: (202)-357-1350

The Smithsonian Associates Research Expedition Program is a branch of the Smithsonian National Associate Program (SNAP). The SNAP is a national membership program that provides educational 'hands-on' opportunities for both Smithsonian Associates and the general public. Participants give their time, money, and physical labor to support projects being directed by Smithsonian personnel. The personnel may be scientists, curators, or research associates of the Smithsonian Institution working to create an exhibition, improve the collections or record data at a field site.

Project volunteers or participants usually spend 1-2 weeks working on a given project. The monetary contribution paid by the participant will depend upon the expenses of the expedition. These contributions to help support research programs conducted by the Smithsonian Instituion are tax deductible. The contribution includes all lodging, supplies, and field transportation. Travel to and from the staging site, as well as visas, passports, medical treatment, and other personal living expenses are not covered by the contribution.

To be eligible, participants must be at least 18 and in good health. Participants must also enroll for the duration of the expedition. Academic credit can be arranged for participation on a Smithsonian Research Expedition. Some of the projects have a small number of scholarships available for students and teachers.

In 1990, the Smithsonian Associates Research Expedition Program will have at least two projects being conducted in Latin America. One is the Costa Rican Volcano: Reshaping the Landscape. This is a 12 day expedition that has a staging site in San Jose, Costa Rica and a required contribution of $1,395. The other is Yucatan Birdlife and Vegetation, with a staging site in Cancun, Mexico and a required contribution of $1,450.

University Research Expeditions Program (UREP)
University of California
Berkeley, CA 94720 USA
Telephone: (415)-642-6586

The University Research Expeditions Program (UREP) is a non-profit organization that was established in 1976. It functions to bring the public and the scientific community together and encourage greater communication and sharing of information and skills. It enables the funding of important and exciting research projects which might not otherwise receive traditional support, because a scientist is just starting out or the research is too innovative. UREP matches people from all walks of life with University of California scholars. Participants become an active member of the field team and contribute an equal financial share to cover the project's costs. A participant's donation to the University of California is tax-deductible.

Most UREP projects have groups of four to eight participants assisting at any one time. The field period may last from two to four weeks. UREP administers, coordinates and publicizes the expeditions and prepares informational materials for participants. In the 1989 UREP listing of expeditions, there were four projects taking place in tropical forests: Maquipucuna-Study Of An Ecuadorian Rain Forest, Lomas Barbudal-Development Of A Wildlife Reserve (Costa Rica), Birds Of The Kalamega Rain Forest (Kenya), and Monkeys Of Borneo. The first project required meeting at Quito (Ecuador), a two week field period, and a contribution of $1,265.

A contribution normally covers meals and shared lodging during the dates of an expedition, ground transportation, camping and field gear, research equipment and supplies, information regarding field techniques, recommended reading, project area and conditions, and a list of things you need to bring. The contribution does not include air fare or travel costs to the assembly point; visas, passports, or inoculations, medical treatment or emergency evacuation expenses; personal expenses.

No previous training or experience is required for most UREP expeditions. Team members will work alongside university researchers and their staff, learning new skills on site. Applications and a current list of UREP projects can be obtained from the above address.

CHAPTER 4   SOURCES OF FUNDING

INTRODUCTION

This chapter is directed primarily at those readers involved in academia, with eventual plans of conducting original research. Unless one is independently wealthy and can afford to finance their own research, it will be necessary to write and submit a grant proposal to one of the organizations that provides funding. Most granting agencies require that the applicant be affiliated with a scientific institution or university, but some have programs open to high school students, teachers, and laymen.

Very few funding organizations provide financial support specifically for natural history-oriented research (the Roger Tory Peterson Institute is an exception). Most projects stand a better chance of receiving funds if they involve an organism or theme of economic importance, or if it can be demonstrated that the research may have far reaching effects and be of value to the scientific community in general. Eligibility guidelines and application procedures vary among agencies, while research emphasis and special programs within an organization may change from year to year. It is therefore advisable to solicit a current set of application materials and guidelines from each organization of interest, each time that a proposal is submitted. Granting agencies often take a considerable time to evaluate proposals (six months to a year is not uncommon), and frequently will only accept grant applications at certain times of the year. Both these points are important to remember when planning and scheduling a research project.

The following institutions and organizations listed do not form a definitive list. It is merely meant to serve as an aid to students and others who are beginning the process of trying to obtain information about financial support for biological research. Many universities offer courses in grant writing and have personnel whose function it is to locate potential sources of funding. If a university has a Center for Latin American Studies, this can be an excellent place to inquire about specially funded grants and programs. The Center for Latin American Studies at the University of Pittsburgh for example, put out a booklet titled:

GUIDE TO FINANCIAL AID FOR LATIN AMERICAN STUDIES. This can be obtained from:

> Center for Latin American Studies
> University of Pittsburgh
> 4EO4 Forbes Quadrangle
> Pittsburgh, PA 15260 USA
> Telephone: (412)-648-7392

An additional reference listing sources of financial support, specifically pertaining to Latin America is *FUNDING FOR RESEARCH STUDY AND TRAVEL, LATIN AMERICA AND THE CARIBBEAN*, edited by Karen Cantrell and Denise Wallen. This book can be obtained from:

> The Oryx Press
> 2214 North Central at Encantro
> Phoenix, AZ 85004-1483 USA

The preservation of biological diversity and the conservation of the world's tropical forests are 'hot' topics today, whose importance has been recognized by various private foundations. Many private foundations exist, some of which provide funding for research in biology or have special programs supporting conservation. Various ones with a past history of supporting such work are included in the following list, however directories listing hundreds of foundations can usually be found in public or university libraries. Specialized reference books about such foundations can be purchased through the Foundation Center:

> The Foundation Center
> 79 Fifth Avenue, Dept. GF
> New York, N.Y. 10003 USA
> Telephone: (212)-620-4230

Curatorial fellowships are available from many museums and scientific academies that maintain large reference collections of organisms. Such programs usually allow a candidate to work closely for a specified length of time with the institution's expert or curator of a certain group of plants or animals. Those interested in the taxonomy and systematics of particular organisms are encouraged to contact the various museums and botanical gardens. Directories listing such institutions should be available in the reference section of any large library (see page 282 and pages 373-375).

A novel method used by a number of beginning biologists to obtain field experience at little or no cost, is to work as a resident naturalist at a tourist lodge or biological station. Although the particular arrangement varies, it usually involves someone with a background in biology acting as a guide to groups of tourists that visit the area. In return for this service, the naturalist is usually provided with room and board, and allowed to conduct research when not needed as a guide. Of course, the actual amount of free time such a resident naturalist or guide would have depends on how many people visit that particular facility and how many guides are employed at the same time. However, an arrangement can usually be worked out to the benefit of both the lodge owner and the biologist. It is advisable that anyone interested in working as a resident naturalist actually visit the facility and area before applying for such a position. This will help prevent misconceptions regarding the accomodations, food, climate, habitat, etc.

Two facilities that have employed resident naturalist programs in the past are the Explorer's Inn (located at the Tambopata Wildlife Reserve in southeastern Peru) (see pages 10-16) and the Marenco Biological Station (located near Corcovado National Park in the Osa Peninsula of southern Costa Rica) (see page 185). The proprietor of La Selva (located in eastern Ecuador) (see pages 65-70) has also expressed potential interest in this type of arrangement. With the increase of tourism and tourist facilities in rainforested areas, the resident naturalist option may become more available and inviting to an increasing number of biologists.

Selected List Of Organizations With Programs Potentially Supporting Research In The Tropics

Agency For International Development (AID)
Office of the Science Advisor
Room 320 SA-18
Washington, D.C. 20523-1818 USA
Telephone: (703)-875-4444

In 1981, through a congressional mandate, the Agency for International Development (AID) initiated their Program in Science and Technology Cooperation (PSTC). The purpose of this program is to stimulate new and innovative scientific research on problems that confront developing countries. This program is administered by the Office of the Science Advisor (AID/SCI) and provides highly competitive research grants to qualified applicants.

The PSTC gives highest priority to submissions from scientists in those developing countries which receive USAID development assistance. In Latin America, this refers to Belize, Bolivia, Costa Rica, Ecuador, Guatemala, Honduras, Panama, and Peru. The grants competition is also open to U.S. investigators and scientists from middle income countries (Argentina, Brazil, Colombia, Malaysia, Mexico, Nigeria, and Venezuela). The funding of proposals originating from those countries usually occurs when the research is particularly relevant, unusually innovative, and there is strong scientific collaboration (intellectual partnership) with developing country investigators. The PSTC seeks new research ideas in the natural sciences and engineering. Innovative ideas that will eventually lead to solutions to serious developing country problems are accorded highest priority.

The AID/SCI has identified six areas of investigation for special emphasis and priority funding. These areas are: 1) Diversity of Biological Resources, 2) Biomass Resources and Conversion Technology, 3) Plant Biotechnology, 4) Biological Control, 5) Chemistry for World Food Needs, and 6) Biotechnology/ Immunology. More information about the PSTC and the application process can be obtained by writing to the above address.

American Association of University Women Education Foundation
2401 Virginia Avenue, N.W.
Washington, D.C. 20037 USA
Telephone: (202)-728-7603

The American Association of University Women (AAUW) Education Foundation is the philanthropic arm of the AAUW. Among the programs it administers is the American Fellowships program, for women who are U.S. citizens or permanent residents. Financial aid is provided to women at the dissertation and post-doctoral levels for full time study or research.

Earthwatch/Center For Field Research
680 Mt. Auburn Street
P.O. Box 403
Watertown, MA 02272 USA
Telephone: (617)-926-8200

Earthwatch is a non-profit organization that provides funding and manpower for field research projects (a detailed description of Earthwatch is given on pages 319-320). Funds are obtained from laymen volunteers that pay a pro-rated share of the total expedition costs. They then work as field assistants at the actual research site, side by side with the principal investigator. Scientists wishing to apply for support from Earthwatch should request a set of proposal forms from the Center For Field Research.

The proposal addresses all components of the project, including the various duties to be performed by the volunteers. It is evaluated both by the Center For Field Research and outside peer reviewers. If it is judged to be of sufficient scientific merit with interest enough to attract the necessary volunteers, it will be accepted and advertised in Earthwatch's periodical magazine of expeditions.

For laymen wishing to participate on a particular expedition, Earthwatch has offered various scholarship programs, which subsidized part or all of the expedition contribution for the recipients. One such program was specifically aimed at high school teachers. Earthwatch should be contacted at the above address to determine which special programs are currently in effect. Last year over 250 teachers and students received fellowship help to work on projects.

Explorers Club
46 East 70th Street
New York, N.Y. 10021 USA
Telephone: (212)- 628-8383

The Exploration Fund of the Explorers Club is used to make grants in support of field research and exploration, usually not exceeding $1,000. Applications describing the project and investigator are submitted to an evaluation committee of the Explorers Club. It is judged on the practical and scientific merit of the proposal, whether the budget appears realistic, and the competence of the investigator. Expeditions are aided in accordance with the Explorers Club stated objective, "to increase knowledge of the world." Mere travel to remote areas or big game hunting will not be considered.

Foundation For Field Research
P.O. Box 2010
Alpine, CA   92001-0020   USA
Telephone: (619)-445-9264

The Foundation For Field Research (FFR) is a non-profit organization that provides grants to researchers, usually ranging from $2,000 - $15,000. All funds for grants are derived from share-of-cost contributions by volunteer participants on the research teams. Funding is available to citizens of any nation. Investigators with a strong background in the proposed project's discipline, as demonstrated in their curriculum vitae, are eligible to apply.

In addition to funds, the FFR maintains a large inventory of expedition equipment that is available to researchers. Among the items available are vehicles, inflatable boats with outboard motors, archaeological equipment, wall tents and other materials necessary for establishing base camps in the field. Proposals are accepted for projects any place in the world that does not pose a danger to travelers. Grant applications should be submitted 12-14 months prior to the project starting date. The FFR is also discussed on pages 320-321.

Fulbright Scholar Program
Council for International Exchange of Scholars
3400 International Drive, N.W.
Suite M-500
Washington, D.C.   20008-3097
Telephone: (202)-686-4000

The Fulbright Scholar Program offers grants for faculty and professionals to a large number of Latin American countries. Recipients of these grants may either lecture, conduct research, or a combination of the two. Every March, CIES publishes the annual awards book listing grants, types of awards, countries and subject areas for lecturing and research. Application materials are also available in March.

Two specific programs of interest to tropical biologists that come under the designation of Regional Awards are the American Republics Research Program and the Central American Republics Research Program. Awards for 6 months and 3-6 months, respectively, are given to conduct research in any discipline. In the former program research may be conducted in one or more countries of South America, Mexico, and the Caribbean. In the latter program research may be conducted in one or more countries of Central America. Applications are especially encouraged from scholars whose projects involve collaboration with host country colleagues. Candidates must have a Ph.D. or equivalent professional status, and have sufficient language ability to conduct the proposed research. Benefits, application deadlines, and other information can be obtained by calling the number listed above.

Garden Club of America/World Wildlife Fund
Scholarships in Tropical Botany
1250 Twenty-fourth Street, N.W.
Washington, D.C. 20037 USA
Attn: Jane MacKnight

The Garden Club of America and World Wildlife Fund announced in 1989 the availability of two $5,000 awards to assist with field work in the area of tropical botany. Those eligible are graduate students conducting field work in the tropics as part of their doctoral dissertation research. Awards are competitive and made on a one-time basis. Although U.S. citizenship is not required, students must be enrolled in a U.S. university to be eligible. Specific details for applying can be obtained at the above address.

Man and the Biosphere (MAB) Young Scientists Research Grants
U.S. MAB Secretariat, OES/ENR/MAB
Department of State
Washington, D.C. 20520

These grants are designed to facilitate research work of young scientists in MAB field projects, comparative studies, and biosphere reserves. Applicants should not be older than 35 years. Individual research grants are made up to an amount of $5,000, although smaller requests have greater chances of approval. Candidates from the U.S. may obtain application forms from the U.S. MAB Secretariat at the above address.

McNamara Fellowships Program
The Economic Development Institute
The World Bank
1818 H Street, N.W.
Washington, D.C. 20433 USA

Robert S. McNamara Fellowships are awarded annually to approximately ten scholars or groups of scholars by the World Bank. The purpose of these fellowships is to support innovative and imaginative work that contributes to the general knowledge of economic development. The successful candidate is provided with a stipend for subsistence, travel, accomodations, and an allowance for books and research costs.

There are no restrictions on research topics within the general field of economic development, but each year emphasis is placed on topics in specially selected areas. Basic criteria for applicants are: national of a World Bank member country, normally 35 years old or younger, holds a minimum of a Master's degree or equivalent, work must be carried out in a World Bank member country other than applicant's own. The McNamara Fellowships Program is not intended to support work leading to an advanced degree.

National Geographic Society
Committee for Research and Exploration
17th and M Streets, N.W.
Washington, D.C. 20036 USA
Telephone: (202)-857-7439

The National Geographic Society supports field research all over the world of projects relevant to the field of geography (including anthropology, archaeology, astronomy, biology, botany, ecology, physical and human geography, geology, oceanography, paleontology, zoology, and others). Multidisciplinary and interinstitutional projects are encouraged. Grants are awarded on the basis of the project's scientific merit, with special attention paid to the significance of the proposal, in terms of its relationship to major scientific questions or problems.

Investigators with a doctor's degree and associated with institutions of higher learning, or other scientific and educational non-profit organizations are eligible to apply. In some cases, grants are awarded to exceptionally well qualified graduate students or scientific workers who do not have research degrees or university affiliation. Citizens of any country are eligible.

An information sheet on National Geographic Society research grants, as well as official application forms, are available at the above address.

National Science Foundation
Washington, D.C. 20550 USA

The National Science Foundation (NSF) is a government agency whose main purposes are to increase the nation's base of scientific and engineering knowledge, strengthen its ability to conduct research in all areas of science and engineering, and to develop and implement science and engineering education programs. The foundation does this through merit-selected grants, contracts, and other agreements awarded to universities, university consortia, and non-profit and other research organizations. NSF welcomes proposals on behalf of all qualified scientists, engineers, and science educators, and strongly encourages women, minorities, and disabled individuals to participate fully in its programs.

The National Science Foundation supports many different programs, each with its particular guidelines, deadlines, and eligibility rules. Examples of some of the NSF programs are: Grants For Research In Science And Engineering, Research Experiences For Undergraduates, Fellowship Opportunities In Environmental Biology For New Ph.D.s, Systematic Biology Program, and others. NSF issues detailed brochures announcing and describing new programs, critical dates, and application procedures. Also available is *GRANTS FOR RESEARCH AND EDUCATION IN SCIENCE AND ENGINEERING*, which provides guidance for the preparation of unsolicited proposals to NSF. These are available at the following address:

Forms and Publications Unit, Room 232
National Science Foundation
Washington, D.C. 20550 USA
Telephone: (202)-357-7861

National Wildlife Federation
1412 Sixteenth Street, N.W.
Washington, D.C. 20036-2266 USA
Attn: Environmental Conservation Fellowships
Telephone: (703)-790-4484

The National Wildlife Federation (NWF) is the largest conservation organization in the U.S. It has maintained an Environmental Conservation Fellowship Program since 1956. These fellowships are offered annually to graduate students to encourage advanced study in fields relating to wildlife, natural resource management, and protection of environmental quality.

To be eligible, an individual must be pursuing a degree in a university or college graduate program or law school. The applicant must be enrolled for the Fall semester following the awarding of the grant. Grants up to $10,000 are made for a one-year period, although recipients may apply for additional funding in subsequent years.

The National Wildlife Federation requests proposals that address specific study topics that are of special interest. These topics fall under the broad headings of Wildlife, International, Education, Habitat, Fisheries, Pollution-Toxics-Pesticides, and

Issues-Policies-Legislation. The specific topics pertaining to each heading are available, along with application materials, from the NWF at the above address.

Jessie Smith Noyes Foundation
16 East 34th Street
New York, N.Y. 10016 USA
Telephone: (212)-684-6577

The Jessie Smith Noyes Foundation (JSNF) makes grants to tax-exempt organizations. It has the goal of preventing irreversible damage to the natural resources on which human life depends. To address this goal, the JSNF has made grants in the areas of tropical ecology, sustainable agriculture, water resources, and family planning. The objective under 'tropical ecology' has been to retard the destruction of tropical forests, with focus in the western hemisphere. No grants are made to individuals, and preference is given to Latin American organizations. The Foundation rarely makes grants for research unless they are part of a broader program of institution building and/or action.

Organization for Tropical Studies
Box DM
Duke Station
Durham, N.C. 27706 USA
Telephone: (919)-684-5774

The Organization for Tropical Studies (OTS) is a non-profit consortium that promotes the wise use of natural resources in the tropics, by providing leadership in education and research (a detailed description of OTS is given on pages 322-324). OTS maintains field stations in several locations in Costa Rica, for the purpose of accommodating and aiding researchers. In the past, OTS has made small grants to support research in tropical biology. Preference has usually been given to projects conducted at an OTS field station, and to investigators that are alumni of the OTS field courses offered in Costa Rica. However, this is not a strict rule. Information on current programs implemented by OTS can be obtained at the above address.

Roger Tory Peterson Institute
110 Marvin Parkway
Jamestown, N.Y. 14701 USA
Telephone: (716)-665-2473

The Roger Tory Peterson Institute is an educational organization dedicated to inform society about the natural world through the study and teaching of natural history. Its principal objective is investigation and promotion of the emotional and intellectual linkages of people to nature in ways that promote lifelong curiosity, passion and caring about wild places and things.

Since 1987, the Institute has provided grants ranging to $5,000 for student, amateur and professional study of the natural world, and development of innovative nature education projects. During 1989, grant projects focused on international teacher training and field work involving assessment of the emotional responses of people and cultures to their natural environment. Application guidelines for support of studies of people and nature in the tropics will be available during the first quarter of 1990 and following years.

School For Field Studies
16 Broadway
Beverly, MA 01915 USA
Telephone: (617)-927-7777

The School for Field Studies (SFS) is a unique educational organization that offers field-based research-oriented courses of one month or one semester (13 weeks) duration. These courses investigate a wide variety of ecosystems and environmental issues, and are designed to instruct students in the mastery of field techniques while living in the ecosystem under study. All SFS courses are accredited through Northeastern University in Boston. SFS is the largest private institution in the U.S. devoted exclusively to formal coursework and field study in conservation biology.

The School for Field Studies maintains four research centers in different parts of the world, with a Center for Rainforest Studies located in Queensland, Australia. Mobile bases located in the neotropical forests of Costa Rica, Ecuador, and Mexico have recently hosted one month long SFS summer courses on tropical ecology, ethnobotany, and experimental social ecology, respectively. Tuition for the previous three courses was approximately $2,000 each, plus a $100 registration fee. Participants must also provide their own transportation to the designated course departure site.

The School for Field Studies provides financial aid based on need, in the form of scholarships and interest-free loans to students participating in SFS courses. Interested students who will need financial aid in order to participate are encouraged to submit a SFS financial aid application, which can be obtained from the SFS Admissions Office. Applicants for admission to the month-long courses need to have completed at least their junior year of high school.

Sigma Xi
The Scientific Research Society
Committee on Grants-in-Aid of Research
345 Whitney Avenue
New Haven, CN 06511 USA
Telephone: (203)-624-9883

The Sigma Xi Society makes awards to support research in any field of scientific investigation. Awards are made payable to individual recipients and do not normally exceed $600. Priority is usually given to applicants who are in an early stage of their scientific careers. Applications can be obtained from the above address and are evaluated at three times of the year.

Smithsonian Institution
Office of Fellowships and Grants
Suite 7300
955 L'Enfant Plaza
Washington, D.C. 20560 USA
Telephone: (202)-287-3271

The Smithsonian Institution offers a variety of funding programs for investigators interested in conducting research at one of their facilities. Among the facilities of most interest to tropical biologists are the National Museum of Natural History and the Smithsonian Tropical Research Institute (STRI), the latter of which is described in detail on pages 198-203. Examples of

Smithsonian academic programs are internships and predoctoral, postdoctoral, and graduate student fellowships. Descriptions of current programs are provided in the publication *SMITHSONIAN OPPORTUNITIES FOR RESEARCH AND STUDY*, which also describes each Smithsonian facility, its respective research staff, and the staff's areas of expertise. This book and application materials may be obtained from the Office of Fellowships and Grants.

Research Development Awards are also offered through the National Zoological Park. Information on this program can be obtained by writing to the following address:

Assistant Director for Research
National Zoological Park
Washington, D.C. 20008

The Tides Foundation
1388 Sutter Street, 10th Floor
San Francisco, CA 94109 USA
Telephone: (415)-771-4308

The Tides Foundation was created in 1976 to promote creative non-profit and philanthropic activity. Its original focus was in the western U.S., but it has since expanded its scope to national and international regions. The Tides Foundation supports efforts in the following five general areas: 1) Land Use, Preservation and Stewardship, 2) Economic Public Policy and Enterprise Development, 3) Environment and Natural Resources, 4) International Affairs, 5) Community Affairs and Civil Liberties. A main theme of the first area mentioned is the preservation of the planet's remaining wildlands and rainforests. A project of the Tides Foundation is the International Rivers Network (see page 297).

The Tides Foundation makes grants to the following categories of recipients: 1) To individuals directly for charitable and educational purposes or in recognition of past achievements in religious, arts, charitable, scientific, educational or civic work, 2) To organizations exempt from federal income tax, and 3) To other organizations exclusively for charitable or educational purposes. The Foundation evaluates both solicited and unsolicited proposals. It requests that prospective applicants submit only a two page letter of inquiry explaining: 1) The issue being addressed, 2) The organization's approach to the issue, 3) The audience for the project, and 4) The budget requirements and funding received to date. If the original inquiry passes the initial review process, the Tides Foundation may request a full proposal. Inquiries can be sent to the above address.

University Research Expeditions Program (UREP)
University of California
Desk D-14
Berkeley, CA 94720 USA
Telephone: (415)-642-6586

The University Research Expeditions Program (UREP) provides funds and field assistance to University of California researchers worldwide. Support is provided by selected members of the public who subsidize the research costs through their tax-deductible donations and contribute their own skills and time as short-term field assistants. UREP funds can be used for short- or long-term field research, as seed money for new research, to continue on-going projects, to supplement other grants, and to support graduate students or additional staff.

Faculty or staff researchers from any U.C. campus are encouraged to apply for full or partial funding through UREP. Principal Investigator status is not required for consideration. Graduate students may apply as independent field directors with the sponsorshop of a faculty advisor or receive full or partial funding as assistants to a faculty member.

Projects in any discipline which can incorporate UREP volunteer participation will be considered. Research projects involving specimen or artifact collection, excavation, interviewing, field observations, photography, surveying, etc. are particularly suited to UREP funding. For applications and proposal deadlines, as well as information about the review and evaluation procedures for proposals, please write to the above address.

UREP also offers a Grants for Teachers Program, with funds used to cover a portion of the expedition costs, and in some cases, a portion of the air fare to the project area. Teachers in natural or social sciences are eligible, with California teachers in grades K-12 given priority. Others will be considered on a space available basis. For further information and application materials, write:

> University Research Expeditions Program (UREP)
> Teacher Program, Desk D02
> University of California
> Berkeley, CA 94720

A limited number of partial scholarshops are available to students participating in UREP expeditions. Interested students should contact the UREP office for special student scholarship applications.

Whitehall Foundation
Suite 202
249 Royal Palm Way
Palm Beach, FL 33480 USA
Telephone: (407)-655-4474

The Whitehall Foundation assists research in the life sciences through its programs of grants and grants-in-aid. Its policy is to assist the areas of basic biological research that are not otherwise heavily supported. Although consideration is given to applicants regardless of age, preference is given to scientists at the beginning of their careers and many senior scientists. Research grants normally range from $10,000 to $40,000 per year, while grants-in-aid are limited to $15,000. One should contact the Whitehall Foundation at the above address for their current areas of interest and application procedures.

Wildlife Conservation International
New York Zoological Society
Bronx, NY 10460 USA
Telephone: (212)-220-5090

Wildlife Conservation International (WCI) provides grants to individuals affiliated with an institution or an institution itself, to support proposals addressing specific conservation concerns for a particular species or assemblage of species, or more broadly conceived topics in conservation biology. In the recent past grants have ranged from $150 to $100,000, averaging under $20,000. All proposals are evaluated on a competitive basis with the WCI Conservation Committee making final selections. The Conservation Committee meets twice a year and the closing dates for applications are January 1 and July 1. Grants are awarded by June and November of each year.

Application materials may be obtained from WCI at the above address. It is strongly suggested that a brief letter of inquiry be sent before applying. Completed applications are sent to the attention of Dr. Mary Pearl. The following are not considered priority conservation activities by WCI: airfares to scientific meetings, support of conferences, legal actions, and erection of permanent field stations. Stipends are only considered when the investigator has no other source of support. Included among the responsibilities of successful applicants to host countries are: training nationals in wildlife research, keeping local and national officials informed of project activities and results, and lecturing at local schools and other institutions.

World Wildlife Fund
Biodiversity Program Director
1250 24th Street NW
Washington, D.C. 20037 USA
Telephone: (202)-778-9600

World Wildlife Fund is the lead organization in a joint venture with the WRI Center for International Development and Environment (CIDE) and The Nature Conservancy (TNC) to implement a U.S. Agency for International Development (USAID) centrally-funded program for the Conservation of Biological Diversity. The goal is to promote sustainable economic development in developing countries through better conservation and use of biological resources. Initial objectives are to improve the capability of selected developing countries and AID assistance programs to identify critical needs for realizing the economic potentials of conserving and wisely managing biological resources, protecting ecological processes, and maintaining genetic diversity.

The project has five major components: 1) Technical assistance offered to AID missions, host country institutions, local PVOs and the U.S. Peace Corps, 2) A small grants program for specific research issues relevant to USAID's conservation activities worldwide, 3) Training for local scientists in identifying research priorities and preparing competitive proposals, 4) An information and evaluation network on USAID conservation activities linked with other relevant data bases, and 5) Pilot demonstration projects funded largely by USAID mission and regional bureau buy-ins.

The project works closely with the Consultative Group on Biological Diversity that includes several private foundations with active portfolios of conservation projects. For further information on the Conservation of Biological Diveristy Program, please contact Dr. Gary S. Hartshorn at the above address.

An excellent pamphlet listing and describing various grants and fellowships is *A SELECTED LIST OF FELLOWSHIP OPPORTUNITIES AND AIDS TO ADVANCED EDUCATION*. This booklet can be obtained from:

> The Publications Office
> National Science Foundation
> 1800 G Street, NW
> Washington, D.C. 20550 USA
> Telephone: (202)-357-7861

APPENDIX A   Traveling And Travel Agencies

There are many options available for visiting a rainforest, depending on one's reasons for traveling. A tourist traveling to the jungle for the first time may feel more comfortable on a package tour, where the itinerary is pre-planned and all logistical details are taken care of by the tour agency. There are many companies today that offer such natural history tours throughout Central and South America. The tour may be designed to give its participants a general overview of the rainforest flora and fauna, or in other cases it may be developed specifically for a group with similar interests (most often birdwatching).

Accomodations on such tours may vary greatly. One company may offer a wilderness experience that involves camping, while another may specialize in cruises along jungle rivers. Most involve spending some time at a tourist lodge or hotel, constructed in or near the rainforest. I have tried to cover the majority of such facilities used by major tour operators in Chapter 1, so the reader will have a clear idea of what specific lodges are like. Misconceptions as to what a trip will be like often result in a disappointing time. One should try to find out as much as possible about the area to which they will be traveling.

Certain scientific institutions and conservation organizations sponsor natural history trips, usually guided by one of their staff scientists. This is typical of most of the major museums, zoos, and some botanical gardens. Such institutions will sometimes give a travelog or slide show of their trips at certain times of the year in order to generate interest among the public. Interested parties wishing to know more about such programs should call the appropriate institution or organization.

For those readers who wish to have a closer, more scientific look at the rainforest, I suggest reviewing Chapter 3 ('Hands-On' Organizations). It describes programs that enable both laymen and students to participate in on-site research or education in some aspect of tropical biology.

The following is a partial list of the many companies and institutions that offer natural history or special interest tours to rainforest areas. Advertisements for such companies can usually be found in the back of natural history and travel magazines. One should select a tour and tour company carefully, making sure that the selected trip will accomplish the goal of the traveler. For example, if your reason for making a trip is to see a jaguar, you should be aware that although almost every tour brochure mentions jaguars, your chances of actually seeing one are extremely slim (except under very special circumstances in very select locations).

For those who would like to know if traveling to specific countries poses any risks, I suggest calling the Citizens' Emergency Center of the State Department. The telephone number is: (202)-647-5226.

Note: The appearance of the name of any company and/or institution on this list is not to be considered as an endorsement by the author. This list is provided merely to supply a starting point for those readers interested in seeking travel information.

Adventure Associates
13150 Coit Road
Suite 110
Dallas, TX 75240 USA
Telephone: (800)-527-2500 or (214)-907-0414

Adventures Unlimited
P.O. Box 22
Stelle, IL 60919 USA
Telephone: (815)-253-6390

American Museum of Natural History
Discovery Cruises
Central Park West at 79th Street
New York, N.Y. 10024-5192 USA
Telephone: (800)-462-8687 or (212)-769-5700

Basic Foundation
P.O. Box 47012
St. Petersburg, FL 33743 USA
Telephone: (813)-526-9562

APPENDIX A

Betchart Expeditions Inc.
21601 Stevens Creek Boulevard
Cupertino, CA 95014 USA
Telephone: (408)-252-4910
(800)-252-4910

Budget Birding
1731 Hatcher Crescent
Ann Arbor, MI 48103 USA
Telephone: (313)-995-4357

Caligo Ventures
387 Main Street
P.O. Box 21
Armonk, NY 10504-0021 USA
Telephone: (914)-273-6333
(800)-426-7781

Canyon Explorers Club
1223 Frances Avenue
Fullerton, CA 92631 USA

Exploration Holidays
A Division of Exploration Cruise Lines, Inc.
1500 Metropolitan Park Building
Seattle, WA 98101 USA

Field Guides Incorporated
P.O. Box 160723
Austin, TX 78746 USA
Telephone: (512)-327-4953

Force 10 Expeditions Ltd.
P.O. Box 547
New Canaan, CT 06840 USA
Telephone: (203)-966-2691
(800)-888-9400

Forum Travel International
91 Gregory Lane, #21
Pleasant Hill, CA 94523 USA
Telephone: (415)-946-1500 or (415)-671-2900

Great Expeditions Tours
5915 West Boulevard
Vancouver, British Columbia
Canada V6M 3X1
Telephone: (604)-263-1505

Holbrook Travel, Inc.
3540 N.W. 13th Street
Gainesville, FL 32609 USA
Telephone: (904)-377-7111

International Expeditions
1776 Independence Court
Suite 104
Birmingham, AL 35216 USA
Telephone: (800)-633-4734

International Zoological Expeditions, Inc.
210 Washington Street
Sherborn, MA 01770 USA
Telephone: (617)-655-1461

Joseph Van Os Nature Tours, Inc.
P.O. Box 655
Vashon Island, WA 98070 USA
Telephone: (206)-463-5383

Journeys
4011 Jackson
Ann Arbor, MI 48103 USA
Telephone: (313)-665-4407

Learning Alliance
494 Broadway
New York, NY 10012 USA
Telephone: (212)-226-7171

Lost World Adventures
1189 Autumn Ridge Drive
Marietta, GA 30066 USA
Telephone: (404)-971-8586

Massachusetts Audubon Society
Natural History Travel Program
South Great Road
Lincoln, MA 01773 USA
Telephone: (800)-289-9504

McHugh Ornithology Tours
101 W. Upland Road
Ithaca, N.Y. 14850 USA
Telephone: (607)-257-7829

Mountain Travel
6420 Fairmount Avenue
El Cerrito, CA 94530 USA
Telephone: (800)-227-2384 or (415)-527-8100

Nature Conservancy
International Trips Program
1785 Massachusetts Avenue, N.W.
Washington, D.C. 20036 USA
Telephone: (202)-483-0231
    (administered in 1989 by International Expeditions, Inc.)

Nature Expeditions International
474 Willamette Street
P.O. Box 11496
Eugene, OR 97440 USA
Telephone: (503)-484-6529

New England Tropical Forest Project
P.O. Box 73
Strafford, VT 05072 USA
Telephone: (802)-765-4337

New York Botanical Garden
Bronx, N.Y. 10458-5126
Telephone: (212)-220-8700

Overseas Adventure Travel
349 Broadway
Cambridge, MA 02139 USA
Telephone: (800)-221-0814 or (617)-876-0533

Questers
Worldwide Nature Tours
257 Park Avenue South
New York, N.Y. 10010 USA
Telephone: (212)-673-3120

Safaricentre
3201 N. Sepulveda Blvd.
Manhattan Beach, CA 90266 USA
Telephone: (800)-223-6046 or (213)-451-2900

Smithsonian Associates Travel Program
Smithsonian Institution
1100 Jefferson Drive, S.W.
Washington, D.C. 20560 USA
Telephone: (202)-357-4700

Society Expedition Cruises
3131 Elliot Avenue
Suite 700
Seattle, WA 98121 USA

or contact

Golden Bear Travel
2171 Francisco Blvd.
San Rafael, CA 94901 USA
Telephone: (800)-551-1000 (outside CA)

or

(800)-451-8572 (inside CA)

Trailblazer
Explorers and Expedition Outfitters
1321 U.S. Highway 19 South
Suite 505
Clearwater, FL 34624 USA
Telephone: (813)-536-1114

Travel Plans International
1200 Harger Road
P.O. Box 3875
Oak Brook, IL 60521 USA
Telephone: (800)-323-7600 or (312)-573-1400

APPENDIX A

Victor Emanuel Nature Tours, Inc.
P.O. Box 33008
Austin, TX 78764 USA
Telephone: (800)-328-VENT or (512)-328-5221

Wilderness Travel
801 Allston Way
Berkeley, CA 94710 USA
Telephone: (800)-247-6700 or (415)-548-0420 (inside CA)

Wings, Inc.
P.O. Box 31930
Tucson, AZ 85751 USA
Telephone: (602)-749-1967

Wonder Bird Tours
200 Fifth Avenue
New York, NY 10010 USA
Telephone: (212)-840-5961

World Wildlife Fund
1250 Twenty-Fourth Street, N.W.
Washington, D.C. 20037 USA
Telephone: (202)-778-9548

Tips For Travelers

1. Exchange Some Money Before You Arrive At Your Destination. Although you can usually obtain a better exchange rate once you arrive, it can be very convenient for taxis, tips, etc. to have some of the correct currency in-hand when you get there. Most major airports have money exchange booths or *cambios* that will charge a small fee for exchanging currency.

2. Wear A Money Belt. I always wear a leather money belt with a zipper in the back as my 'normal' belt when I travel. It easily holds 12 carefully folded $100 bills and my spare suitcase key. In this way you always know where your spare money is, and in the unfortunate event that you should get robbed, the thieves probably will not take your belt. Items like money belts and leg pouches can be purchased from specialty luggage stores.

3. Travel Light. This has always seemed like good advice to me, although between camera gear and insect collecting equipment, I have never been able to follow it. A good rule of thumb is not to take more luggage than it is possible for you to carry by yourself.

4. Travel With A Companion. This becomes especially improtant if you can't travel light. Aside from making traveling generally more pleasurable and saving money by splitting costs, traveling with someone will make things much, much easier (provided you get along with that someone!). There will be times when you have to look for a taxi, or pay an exit tax, or go to the bathroom, and with someone trustworthy standing by to watch your luggage or help you out, these events don't become worrisome experiences.

5. Always Keep Aside Enough Money To Pay Your Exit Tax. In almost every country you leave, you will have to pay an exit tax at the airport. In Latin America the amount is usually reasonable (between $5 and $10), but it can be as high as $30. You will not be allowed to board your aircraft without proof of having paid your exit tax.

6. Make Copies Of Your Passport And Visa. In the event that your passport is lost or stolen, a photocopy of it may greatly facilitate the actions needed to get a temporary or replacement. I always keep a photocopy of my passport in each piece of my luggage, with a note offering a reward (amount unspecified) if the bag is lost and then returned.

7. Hand Carry Valuables. Expect unlocked baggage (and even poorly secured items) to be rifled. Do not put your jewelry, money, camera, etc. in your suitcases if it is possible for you to hand carry it. For photographers, this certainly applies to all the film you've shot during a trip.

8. Divide Clothes And Equipment Evenly. If you take two or more bags and try to divide up your clothes (and other things) evenly, you will not be devastated if one bag is lost or does not arrive with the others. In addition, it is always prudent to hand carry at least one spare change of clothes and some toilet articles, in case your luggage is lost or your flight delayed.

9. Carry A Small Knapsack. This will enable you to store a change of clothes, toothbrush, book, and whatever other items you feel might come in handy.

10. Always Carry Tape. I have found that a roll of heavy masking tape or duct tape is almost indispensable. String is a close second. If a suitcase or camera case breaks (or is overpacked), tape can help to avoid a disastrous situation.

11. Carry Film In Baggies. If you don't want your film to go through security x-ray machines, have it easily accessible in clear plastic baggies. Upon leaving a country, I always remove the protective film canisters so the security personnel can easily see the film cans.

12. Carry Baggies Or A Waterproof Pouch. This becomes important once you reach the rainforest, or when you are traveling by river. It is often very hunid in the tropics with the result that things that get wet tend to stay wet. I always carry a large baggie in a pocket that I can slip my camera into if it starts to rain.

13. Use Repellant With Highest Percentage Of DEET. The active ingredient of most modern insect repellants is a chemical listed as DEET. The most effective repellants usually contain the highest percentage of DEET among the active ingredients listed on the side of the container.

14. Bring A Book. Some of my most frustrating and intolerable situations (such as a 12 hour flight delay) have been made barely tolerable by having a book along. (This is also an excellent custom to practice at banks and in college registration lines.)

15. Bring Some Snack Food. Although many of the jungle lodges I have stayed at have served excellent food, some of them have had fairly rigid dining schedules. When you have to wait seven hours between lunch and dinner, some peanut butter and crackers make an excellent 'happy hour' treat. Trail mix or a candy bar will also come in handy if you are out hiking through the forest all day.

16. Inquire About Rates Before Phoning Long Distance From A Hotel. It is almost always much cheaper to speak with someone overseas by having them call you at a prearranged time, or by calling collect. Several times I have made phone calls of 5-10 minute durations from a hotel in Latin America to discover at the time of checking out that they cost more than the entire hotel bill.

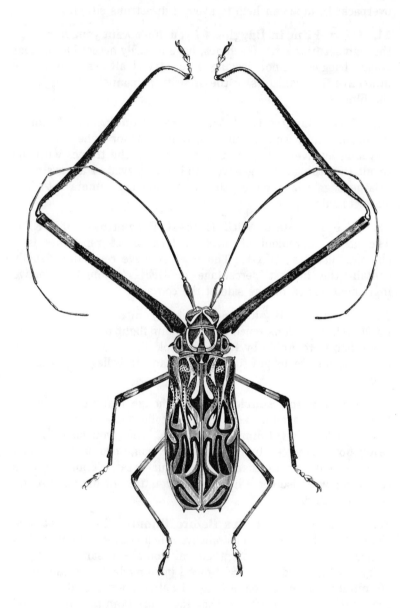

APPENDIX B Specialized Vocabulary (English/Spanish)

Botanical Words

berry	la baya
botanist	el botánico
botany	botanica
bromeliad	la bromelia
bud	la yema
bush	el arbusto
canopy	la copa
epiphyte	la epífita
fern	el helecho
flower	la flor
to flower	florecer
forest	el bosque
cloud forest	el bosque nublado
fruit	la fruta
fungus	el hongo
jungle	la selva
leaf	la hoja
lichen	el liquen
log	el tronco
lumber	la madera
mold	el moho
moss	el musgo
mushroom	el hongo
nectar	néctar
orchid	la orquídea
photosynthesis	fotosíntesis
plant	la planta
pollen	el polen
rainforest	el bosque húmedo

root	la raiz
seed	la semilla
shrub	el arbusto
soil	el suelo
stem	el tallo
sun	el sol
sunlight	la luz solar
tendril	el zarcillo
timber	los árboles
tree	el árbol
trunk	el tronco
vein	la vena
vine	el vejuco o la liana
weed	la maleza, la mala hierba
wood	la madera

Zoological Words

amphibian	el anfibio
bat	el murciélago
biology	la biología
bird	el ave
cat	el gato
cow	la vaca
ecology	la ecología
fish	el pez
frog	la rana
goat	la cabra
horse	el caballo
insect	el insecto
jaguar	el tigre, el jaguar
lizard	la lagartija, el lagarto
macaw	el guacamayo

mammal	el mamífero
monkey	el mono
mouse	el raton
nature	la naturaleza
oscelot	el tigrillo
parrot	el loro, el papagayo
rat	la rata
sloth	el perezoso
snail	el caracol
snake	la serpiente, la culebra
turtle	la tortuga
wildlife	la vida silvestre

Entomological Words

adult	el adulto
ant	la hormiga
bee	la abeja
beetle	el escarabajo
butterfly	la mariposa
caterpillar	la oruga, el gusano
cicada	la chicharra
cockroach	la cucaracha
cocoon	el capullo
cricket	el grillo
dragonfly	el caballito, la libélula
egg	el huevo
fly	la mosca
gnat	el mosquito
to fly	volar
grasshopper	el saltamonte
immature	inmaduro/a
insect	el insecto

katydid	el grillo
larva	la larva
mantid	el mántido
mosquito	el zancudo
moth	la polilla
nest	el nido
nymph	la ninfa
parasite	el parásito
pupa	la crisálida
spider	la araña
tarantula	la tarántula
termite	la termita, el comején
wasp	la avispa

Geographical/Geological Words

bay	la bahía
cliff	el acantilado, el barranco
earthquake	el terremoto
east	este, oriente
the east	el Este, el Oriente
eastern	oriental
flood	la inundación
glacier	el glaciar
hill	la colina, el cerro
ice	el hielo
island	la isla
isthmus	el istmo
key	el cayo
lake	el lago
landslide	el deslave, el derrumbe
mountain	la montaña

north	norte
the north	el Norte
northern	norteño, septentrional
ocean	el océano
oxbow lake	la cocha
peak	la cumbre, el pico
peninsula	la península
plains	los llanos
plateau	la meseta
pond	el charco
rain	la lluvia
to rain	llover
rapids	los rápidos
reef	el arrecife
ridge	la cresta, la cordillera
river	el río
rock	la roca
sea	el mar
slope	el vertiente, la pendiente
snow	la nieve
south	sur
the south	el Sur
southern	meridional, sureño
spring	la fuente, el manantial
strait	el estrecho
stream	el riachuelo
summit	la cima, la cumbre
water	el agua
waterfall	la cascada, la catarata
weather	el tiempo
west	oeste, occidente
the west	el Oeste, el Occidente
western	occidental
wind	el viento

APPENDIX C Tropical Biologists

In addition to the names listed in this appendix, readers might also wish to consult *A WORLD CENSUS OF TROPICAL ECOLOGISTS* (see page 260) and *THE NATURALISTS' DIRECTORY AND ALMANAC (INTERNATIONAL)*, 45th Edition (1989), compiled by Ross H. and Mary E. Arnett.

ORNITHOLOGISTS

Rob Bierregaard
National Museum of Natural History
Stop 106
Smithsonian Institution
10th and Constitution Avenues, N.W.
Washington, DC 20560 USA

Charles Munn III
Wildlife Conservation International
New York Zoological Society
Bronx, NY 10460 USA

Theodore A. Parker III
Museum of Zoology
Louisiana State University
Baton Rouge, LA 70893 USA

Robert S. Ridgely
Academy of Natural Science
19th and the Parkway
Philadelphia, PA 19103 USA

Stuart Strahl
Wildlife Conservation International
New York Zoological Society
Bronx, NY 10460 USA

MAMMOLOGISTS

John F. Eisenberg
Florida Museum of Natural History
University of Florida
Gainesville, FL 32611 USA

Louise Emmons
National Museum of Natural History
Department of Vertebrate Zoology
10th and Constitution Avenue, N.W.
Washington, DC 20560 USA

George Schaller
Director
Wildlife Conservation International
New York Zoological Society
Bronx, NY 10460 USA

John W. Terborgh
Department of Biology
Princeton University
Princeton, NJ 08544 USA

Merlin D. Tuttle
Executive Director
Bat Conservation International
P.O. Box 162603
Austin, TX 78716 USA

HERPETOLOGISTS

Martha L. Crump
Department of Zoology
University of Florida
Gainesville, FL 32611 USA

William E. Duellman
Museum of Natural History
University of Kansas
Lawrence, KS 66045 USA

Roy W. McDiarmid
Biological Survey Section
National Museum of Natural History
U.S. Fish and Wildlife Service
Washington, DC 20560 USA

Charles W. Myers
Department of Herpetology
American Museum of Natural History
Central Park West and 79th Street
New York, NY 10024 USA

A. Stanley Rand
Smithsonian Tropical Research Institute
P.O. Box 2072
Balboa, Republic of Panama

Jay M. Savage
Biology Department
University of Miami
P.O. Box 249118
Coral Gables, FL 33124 USA

ENTOMOLOGISTS

Keith S. Brown, Jr.
Departamento de Zoologia
Instituto de Biologia
Universidade Estadual de Campinas
C.P. 1170
Campinas, Sao Paulo
Brasil 13.100

Thomas C. Emmel
Department of Zoology
University of Florida
Gainesville, FL 32611 USA

Terry L. Erwin
National Museum of Natural History
Department of entomology, NHB-169
Smithsonian Institution
Washington, DC 20560 USA

Charles L. Hogue
Curator of Entomology
Natural History Museum of Los Angeles
900 Exposition Boulevard
Los Angeles, CA 90007 USA

Gerardo Lamas
Director
Museo de Historia Natural
Avenida Arenales 1256
Apartado 14-0434
Lima-14, Peru

Giovanni Onore
Apartado 1160
Quito, Ecuador

David L. Pearson
Department of Zoology
Arizona State University
Tempe, AZ 85287 USA

Michael H. Robinson
National Zoological Park
Smithsonian Institution
Washington, DC 20008 USA

Allen M. Young
Milwaukee Public Museum
800 West Wells Street
Milwaukee, WI 53233 USA

BOTANISTS

Thomas B. Croat
Missouri Botanical Garden
P.O. Box 299
St. Louis, MO 63166 USA

Robin B. Foster
Department of Botany
Field Museum of Natural History
Chicago, IL 60605 USA

Alwyn H. Gentry
Missouri Botanical Garden
P.O. Box 299
St. Louis, MO 63166 USA

Scott A. Mori
The New York Botanical Garden
Bronx, NY 10458 USA

David Neill
USAID/Quito
Agency for International Development
Washington, DC 20523 USA

Mark Plotkin
Conservation International
1015 18th Street, N.W.
Suite 1002
Washington, DC 20036 USA

Ghillean T. Prance
Herbarium Royal Botanical Gardens Kew
Richmond, Surrey
TW9-3AB
England

Francis Putz
Botany Department
University of Florida
Gainesville, FL 32611 USA

CONSERVATION AND POLICIES

Daniel H. Janzen
Department of Biology
University of Pennsylvania
Philadelphia, PA 19104 USA

Thomas E. Lovejoy
Assistant Secretary for External Affairs
Smithsonian Institution Building
1000 Jefferson Drive, S.W.
Washington, DC 20560 USA

Russell A. Mittermeier
Conservation International
1015 18th Street, N. W.
Suite 1002
Washington, D. C. 20036 USA

Norman Myers
Consultant in Environment and Development
Upper Meadow, Old Road
Headington
Oxford OX3 8SZ
United Kingdom

APPENDIX C

Peter H. Raven
Director
Missouri Botanical Garden
P.O. Box 299
St. Louis, MO 63166 USA

Donald E. Stone
Executive Director
Organization for Tropical Studies
Box DM, Duke Station
Durham, NC 27706 USA

APPENDIX D Selected Public Zoos And Botanical Gardens In The United States

Chicago Zoological Park
(Brookfield Zoo)
3300 Golf Road
Brookfield, IL 60513 USA
Telephone: (312)-242-2630 or (312)-485-0263

Cincinnati Zoological Gardens
Zoological Society of Cincinnati
3400 Vine Street
Cincinnati, OH 45220 USA
Telephone: (513)-281-4701

Houston Zoological Gardens
1513 Outerbelt Drive
(POB 1562)
Houston, TX 77030 USA
Telephone: (713)-523-3211

Lincoln Park Zoological Gardens
2200 North Cannon Drive
Chicago, IL 60614 USA
Telephone: (312)-294-4660

Los Angeles Zoo
5333 Zoo Drive
Los Angeles, CA 90027 USA
Telephone: (213)-666-4650

Metrozoo Miami
12400 S.W. 152nd Street
Miami, FL 33177 USA
Telephone: (305)-251-0401

Missouri Botanical Garden
P. O. Box 299
St. Louis, MO 63166 USA
Telephone: (314)-577-5100

National Aquarium In Baltimore
Pier 3
501 East Pratt Street
Baltimore, MD 21202 USA
Telephone: (301)-576-3800

National Zoological Park
Smithsonian Institution
Washington, DC 20008 USA
Telephone: (202)-673-4717

The New York Botanical Garden
Bronx, N. Y. 10458 USA
Telephone: (212)-220-8700

New York Zoological Park
(Bronx Zoo)
New York Zoological Society
185th Street and Southern Boulevard
New York, NY 10460 USA
Telephone: (212)-220-5100

Panaewa Rainforest Zoo
Department of Parks and Recreation
25 Aupuni Street
Hilo, HI 96720 USA
Telephone: (808)-961-8311

Philadelphia Zoological Gardens
Zoological Society of Philadelphia
34th Street and Girard Avenue
Philadelphia, PA 19104 USA
Telephone: (215)-243-1100

St. Louis Zoological Park
Forest Park
St. Louis, MO 63110 USA
Telephone: (314)-781-0900

APPENDIX D   377

San Antonio Zoological Garden And Aquarium
3903 North St. Mary's
San Antonio, TX 78212 USA
Telephone: (512)-734-7184

San Diego Zoological Garden
Zoological Society of San Diego
Balbao Park
(POB 551)
San Diego, CA 92112-0551 USA
Telephone: (619)-231-1515

Message From The Author

As the author of *RAINFORESTS*, I have tried very hard to provide accurate and useful information to the reader. However, conditions and situations sometimes change very rapidly so that some statements may become incorrect or no longer apply, following the printing of this book. There is also the possibility that typographical errors have been made. For whatever the reason, if a reader finds a statement that is untrue or any other type of error, I encourage them to contact me c/o Feline Press.

I would also be most interested in knowing people's reaction and evaluation of *RAINFORESTS*. Especially with regards to information that the reader feels should have been included, but wasn't. I would like to do anything I can to improve the usefulness of this book during subsequent editions.

Thank you very much for your help.

Sincerely,

James L. Castner

c/o Feline Press
P.O. Box 7219
Gainesville, FL
USA 32605

About The Author

Jim Castner was born and grew up in Maplewood, New Jersey (a suburb of Newark) and about as far from tropical rainforests as you can get. However, parks and fields, railroad easements and then undeveloped countryside provided a multitude of entomological subjects for capture and study. Understanding parents and family encouraged his interest in the natural sciences, often aiding in the construction of cages and observation containers, as well as the collection of various food plants for hungry caterpillars. Frequent childhood trips to rural areas of the state created an appreciation of forests and the wildlife within them.

The author received a B.S. in wildlife biology from Rutgers University and an M.S. and Ph.D. in entomology from the University of Florida. His thesis work on the biological control of an introduced agricultural pest has allowed him to travel throughout Central and South America. He has twice received scholarships allowing him to work as an Earthwatch student volunteer on research projects in Costa Rica and Peru. He recently co-conducted an original study of insects in the Amazon Basin, which was funded by Earthwatch.

The author currently works as a scientific photographer for the Entomology Department of the University of Florida. He intends to continue his study and photography of tropical insects and plants, and hopes to eventually visit all the major rainforested areas of the world. He would ultimately like to obtain a position as a curator in a large natural history museum.